CONVERSATION
PIECES

*Inspirational objects in
UCL's historic collections*

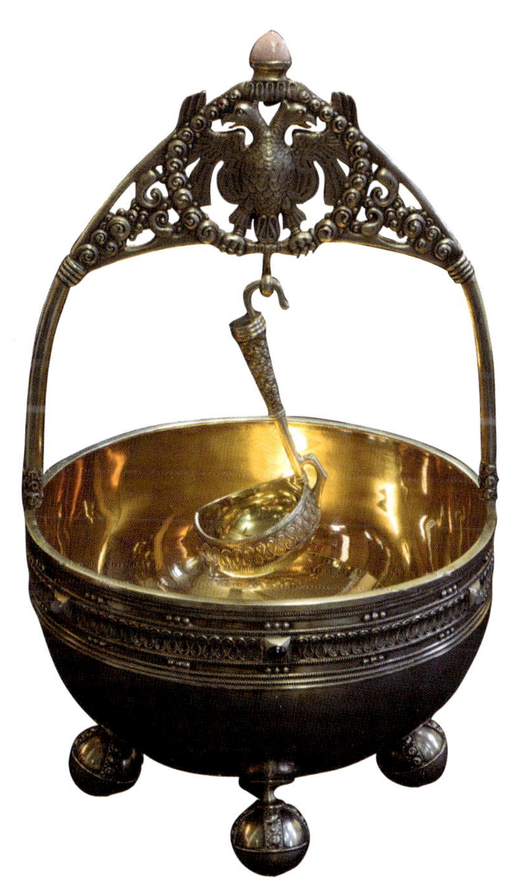

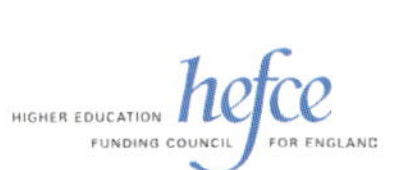

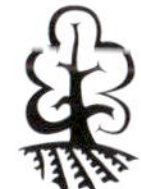

Published in Great Britain in 2013 by Shire Publications Ltd,
PO Box 883, Oxford, OX1 9PL, UK
PO Box 3985, New York, NY 10185-3985, USA

E-mail: shire@shirebooks.co.uk www.shirebooks.co.uk

A CIP catalogue record for this book is available from the British Library.

ISBN-13: 978 1 78200 651 0

Designed by Myriam Bell and typeset in Candara and Garamond Pro.
Editor Ruth Sheppard
Printed in China through Worldprint Ltd.

13 14 15 16 17 16 10 9 8 7 6 5 4 3 2

Front cover image
Blaschka model of squid, *Sepiotuthis sicula*.

Back cover image
Bone calendar or tally device.

Title page image
Pares punch-bowl and salver.

Image Acknowledgments
**Object photography: All images © UCL. Object photography by Colin Dunn of Scriptura: 12,
13, 15, 18, 19, 21, 23, 26, 27, 31 (original image credit: NASA), 34, 35, 38, 39, 43, 46, 49, 55, 57, 58,
59, 62, 63, 66, 67, 69, 71, 72, 73, 75, 79, 81, 82, 83, 85, 86, 87, 91, 94, 95. Mary Hinkley and Tony
Slade UCL Media Services 25, 32, 33, 36, 37, 45, 51, 60, 61, 92.**

**Other Images: © UCL Mary Hinkley UCL Media Services 4. © UCL, Grant Museum of Zoology 5.
© UCL, Grant Museum of Zoology/Matt Clayton 8, 64–5. © Mobile Studios 2011 6, 9. © UCL 7. ©
UCL courtesy Colin Dunn of Scriptura 16–7, 40–1, 96. © UCL, Art Museum/Matt Clayton 28–9.
© Edmund Sumner 52–3, 76, 77. © UCL courtesy Stuart Laidlaw 88–9.**

Preface

Professor Michael Arthur

President and Provost

UCL was the third university to open in England (after Oxford and Cambridge), and the first in England to teach many disciplines, among them Chemistry, Geography and Egyptology. To support teaching and research in these and other subjects, the university quickly began building rich museum and library collections. These now number around a third of a million objects, specimens and artworks, many of them of international significance.

We have wonderful display spaces where these collections are exhibited, but these are also museum laboratories, places of innovation and experiment. If university collections are to remain relevant, they must be active, dynamic; provoking enquiry, stimulating debate, opening minds. From the beginning, UCL's collections were acquired for close study and handling, and were sometimes destroyed in the interests of research. Our most recently acquired collection – the Materials Library in the Institute of Making – is likewise there to be used, but unlike the early collections it is explicitly to be used by all disciplines; engineers, artists, chemists, archaeologists. Increasingly we see a role for our museums not only as specialist resources but as catalysts for new kinds of interaction.

UCL's history of forging international connections means that our museums hold objects from all over the world. This gives us tremendous potential to use our collections as the starting point for international dialogue, whether through repatriation, loans or collaborative exhibitions.

In publishing this book, therefore, we wanted to convey something of the special nature of collections, of the particular relationship that these artefacts have to the academic mission of the university and of the impact they have on our current thinking. A traditional descriptive museum catalogue did not seem appropriate. Instead, we asked a few of our star academics to choose an object – any object – from our extraordinary museums, and to write briefly about its meaning for them, continuing a dialogue between the past and the present. We hope that these 'conversation pieces' will inspire new interactions with our community of staff and students and new relationships with people across the world.

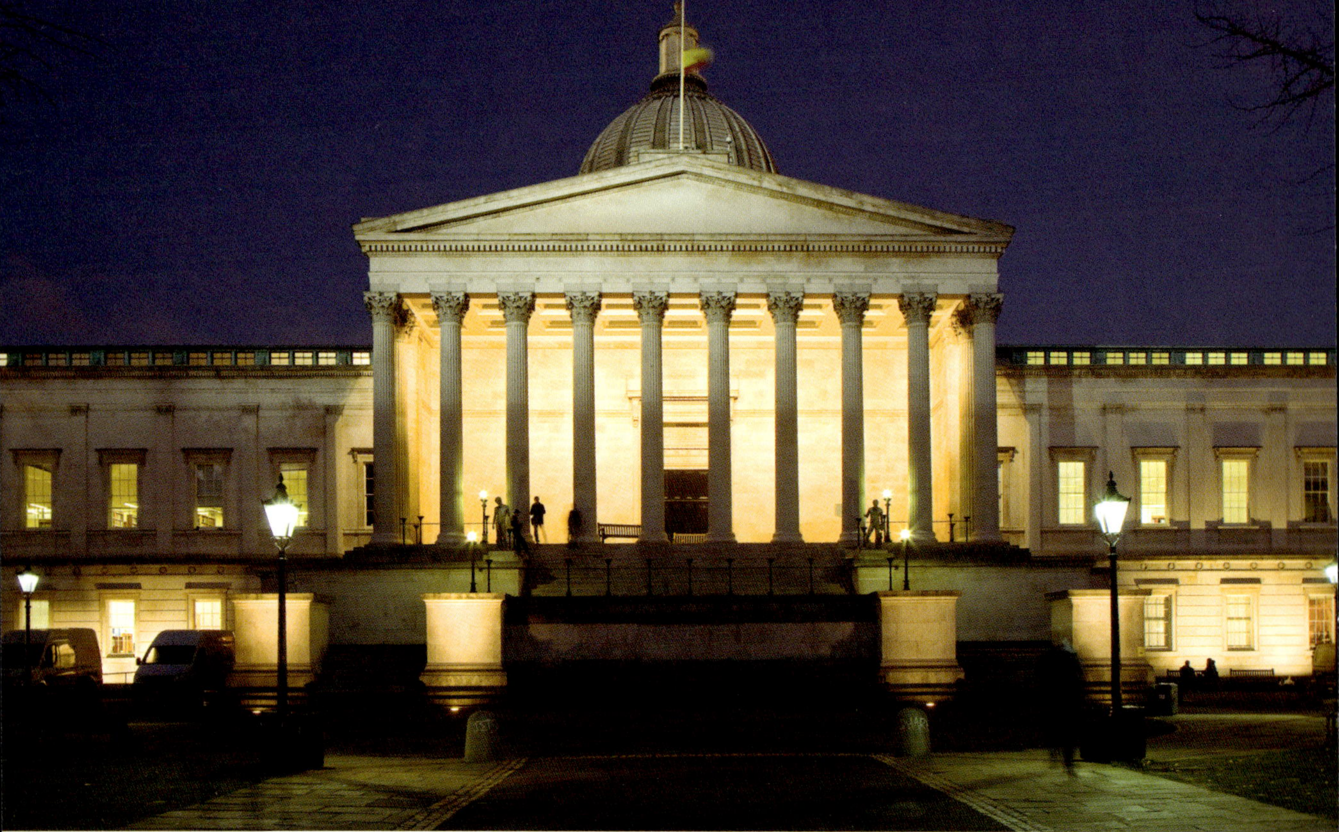

UCL Front Quad and Wilkins Portico.

UCL and the university museum

Sally MacDonald

Director of UCL Museums and Public Engagement

COLLECTIONS HAVE BEEN at the heart of UCL from the very beginning of the university's history. The first *Prospectus*, published in 1826, announced:

The Council are now about to lay the foundations of an institution well adapted to communicate liberal instruction … with the advantage which accrues to all from the outward aids and instruments of Libraries, Museums and Apparatus.

Nineteenth-century plans of UCL show museums of anatomy and natural history in prominent locations near to libraries and laboratories, and early photographs of these museums show objects and specimens out on tables, or in open cabinets. These were clearly not displays simply to be admired; they were essential teaching tools to be used, studied, and sometimes destroyed in the process of enquiry.

UCL was not unusual in this regard. As David Murray commented in 1904: 'Every professor of a branch

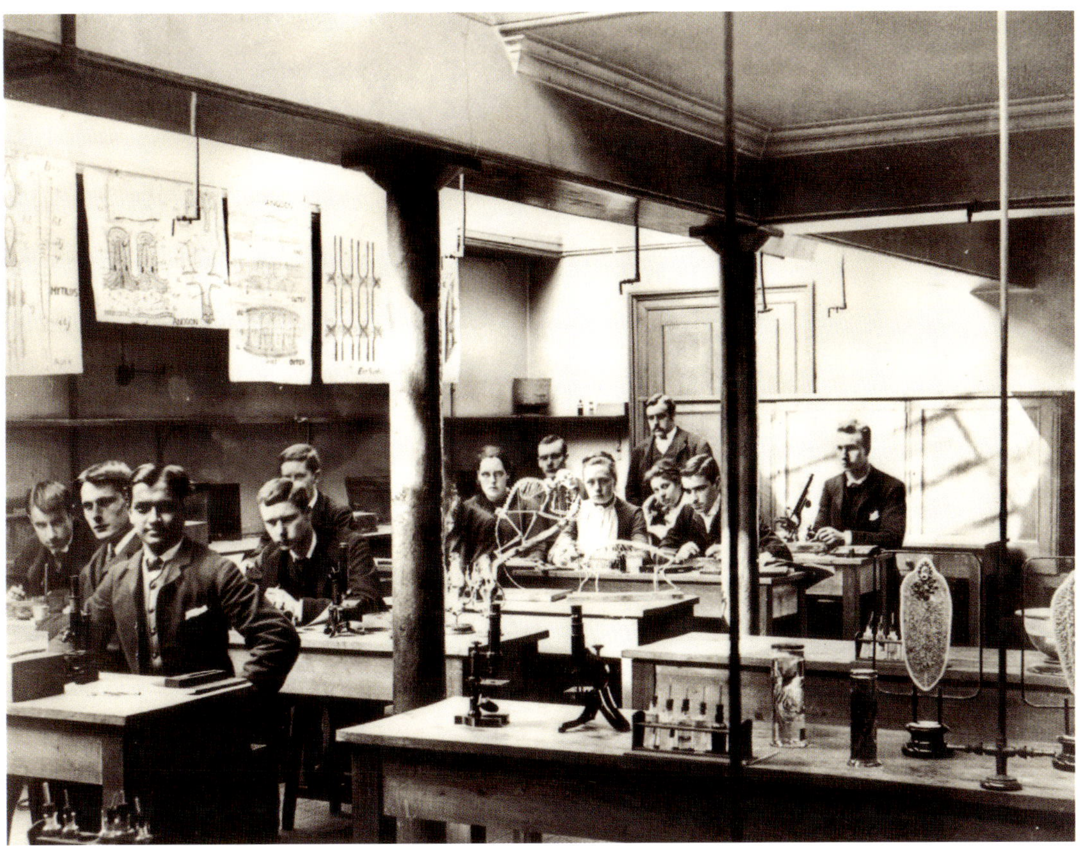

of science requires a museum and a laboratory for his department'. Thus Robert Grant, the radical biologist who joined UCL in 1827 as England's first professor of Comparative Anatomy, immediately began amassing zoological specimens for use in teaching. Flinders Petrie, professor of Egyptology from 1893, brought with him his own collection of ancient Egyptian artefacts, and added to it year on year. Petrie focused on collecting smaller items – such as jewellery – that were easy to use in practical teaching, and things that he himself did not fully understand, as potential research projects for his students. So although the collections established by Grant, Petrie and their colleagues were full of extraordinarily beautiful objects, they were not there for prestige or for show: these were working collections designed to educate and challenge students.

Other university collections emerged from research by academics. Henry Johnston-Lavis, who spent much of his life studying volcanoes in the Mediterranean, left his minerals, rocks and detailed gouache drawings to UCL. Sir Ambrose Fleming, whose invention of the thermionic valve paved the way for the invention of radio, left his experimental valves and detailed papers. Such highly specialist, occasionally maverick collections – the physical traces of experiment and enquiry – thrive particularly well in a university context, where individualism and academic freedom are fostered.

This close association between collections and the individual personalities and passions of their creators is of course not exclusive to university museums, but it is a feature of almost every university collection. At UCL many of these individuals have been world-class academics, their research giving collections a depth and richness unusual in other types of museum.

The fortunes of such collections depend, however, on the reputations of their creators and the currency of their thinking. The apparatus and composite photographs produced by Francis Galton, founder of UCL's Eugenics Laboratory, were locked away for many years after the Second World War, inaccessible even to curators, so uncomfortable was the university with this aspect of its history. At various times in the 20th century, certain collections fell dramatically out of fashion, as teaching requirements changed or those who founded them moved on. In 1928 Sir William Beveridge, then chair of UCL's Estates Committee, and no doubt struggling with competing demands on space, declared 'universities have no business with museums'.

What then is the role of the university museum in the 21st century? Can it still function as an 'aid to liberal instruction' and, if so, how and for whom? There are distinctive aspects of the traditional university teaching collection that still serve a key role today. Firstly, universities hold highly specialist collections and can provide access to specialist knowledge. This is particularly vital at a time when most public museums – in the UK at least – can no longer afford to employ subject experts. Secondly, in an age where many modern displays appear both beautiful and timelessly authoritative, there is surely still a case for the 'working museum' or museum as laboratory: a place for close study, for experiment, for creating new knowledge; a place that is intriguing and often messy. And finally – unlike other museums charged with holding collections in perpetuity for the public – university museums have a particular duty to remain relevant to current research and teaching, even if that means replacing or reconfiguring historic collections.

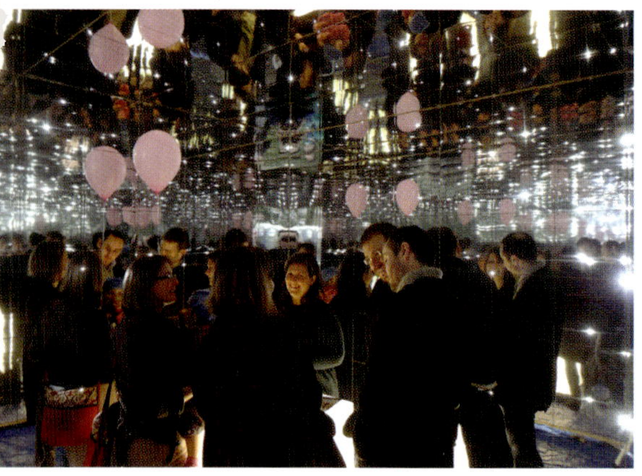

*Intimate encounters inside "The Thing Is",
an experimental space for outreach
with objects.*

Yet if these characteristics are important to retain, there are many other areas where major change is underway. Specialist collections can no longer remain the preserve of individual departments. As universities globally seek to generate new research areas and to equip their students to think laterally, collections increasingly serve as touchstones for cross-disciplinary working. The development and curation of digital – alongside real – collections is essential to support both e-learning and international research. Most significantly of all, many university museums are now not only open to the public but are playing a key role in public engagement and outreach. Visitors to UCL museums today can debate ethical issues with curators via interactive labels, talk to PhD students about their research, test out experimental interpretative systems and encounter pop-up displays by academics.

In UCL's recently opened Institute of Making, a collection is once again at the heart of a new academic initiative. A materials library – including everything from bronze to chocolate – sits alongside a workshop, a 'makespace', in the classic university combination of museum with apparatus. Unlike UCL's founding collections, however, this is a determinedly cross-disciplinary resource, designed to inspire innovation and inventiveness by students of engineering, fine art, psychology or languages.

This book is a celebration of the diversity, range and quality of UCL's collections, which now number around a third of a million objects and specimens. Leading UCL academics from many disciplines were invited to select and write about an object they found inspiring. Their choices were often surprising. Their written contributions are informed, passionate and eloquent. Collectively, they testify to the power of university museums to inspire new generations of students – across disciplines, across continents, across time.

Introduction

Michael Worton

Vice-Provost (International) & Fielden Professor of French Language and Literature

I T IS A STRANGE fact of life that not everyone is excited by museums. Many people visit them only very occasionally, when on holiday, for instance, or when it is raining. Part of the reason for the lack of interest in museums is because they are considered boring and associated with the past. Of course, every museum is in some ways a record of the past – but that is precisely what should make museums interesting. They are places where the history of art, ancient civilizations, animals, fashion, rocks, medicine or whatever is displayed, are interpreted and made into something fascinating and intriguing. This transformation is achieved through the magical work of curators, who, through their choice of objects and the way they talk about them, throw particular light on their collections and draw us into their storytelling as we look at the exhibits. And looking is vital. The famous 19th-century author of *Madame Bovary*, Gustave Flaubert, once said 'Anything becomes interesting if you look at it long enough.' The world is full of fascination – if only we take the time to look at it. And this is where it is important to understand that museums are more than places of the past and of history; they are also places of looking and thinking in the here and now. They are places full of objects which bring with them the wonder and the mystery of half-heard calls from the past combined with interpretations which are very much of the present.

University museums are very special museums, since they are embedded in institutions which are passionate about teaching and research and where academics are constantly creating new knowledge or reviewing, re-assessing and modifying existing knowledge. The university system is an ancient one and it has built up the traditions of different disciplines, which in turn have given rise to the accepted forms of classification and organization in museums. However, in our complex, globalized 21st century, universities are moving away from the established silos of disciplinary work and are engaging in creative interdisciplinarity in teaching and research. And this is increasingly reflected in the way that university museums work, even if they are themselves nominally devoted to specific subjects such as art, archaeology or science.

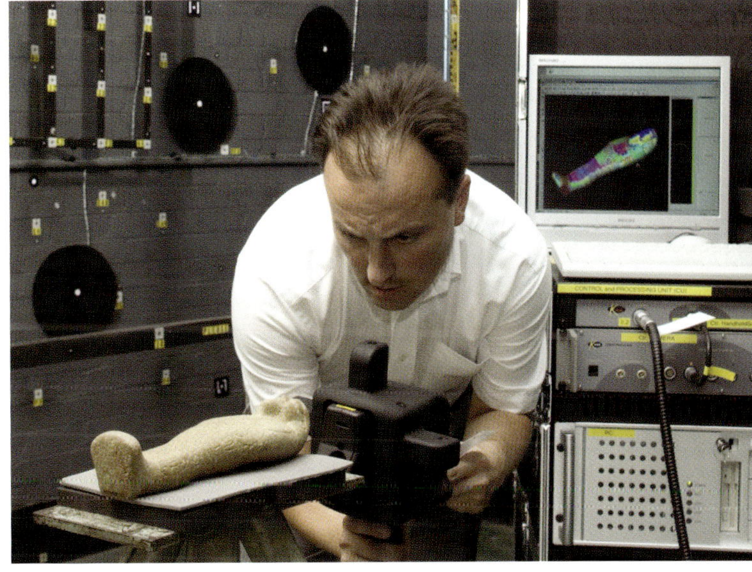

Professor Stuart Robson colour laser scanning an object from the Petrie Museum.

The mission of UCL Museums is a bold one: to redefine the university museum. We are doing this by looking back creatively at the past and looking forward to the future through new attitudes and experimental use of new technologies. For instance, we are pioneering the development of object-based learning. For centuries, artefacts were used in teaching, but then, as two-dimensional, then three-dimensional, and later 3D technologies were developed, objects became less and less used. However, through such scientific approaches as haptics (a feedback technology which studies how the sense of touch works), UCL curators and scientists are exploring how we can learn better through using and touching objects. Here, advances in

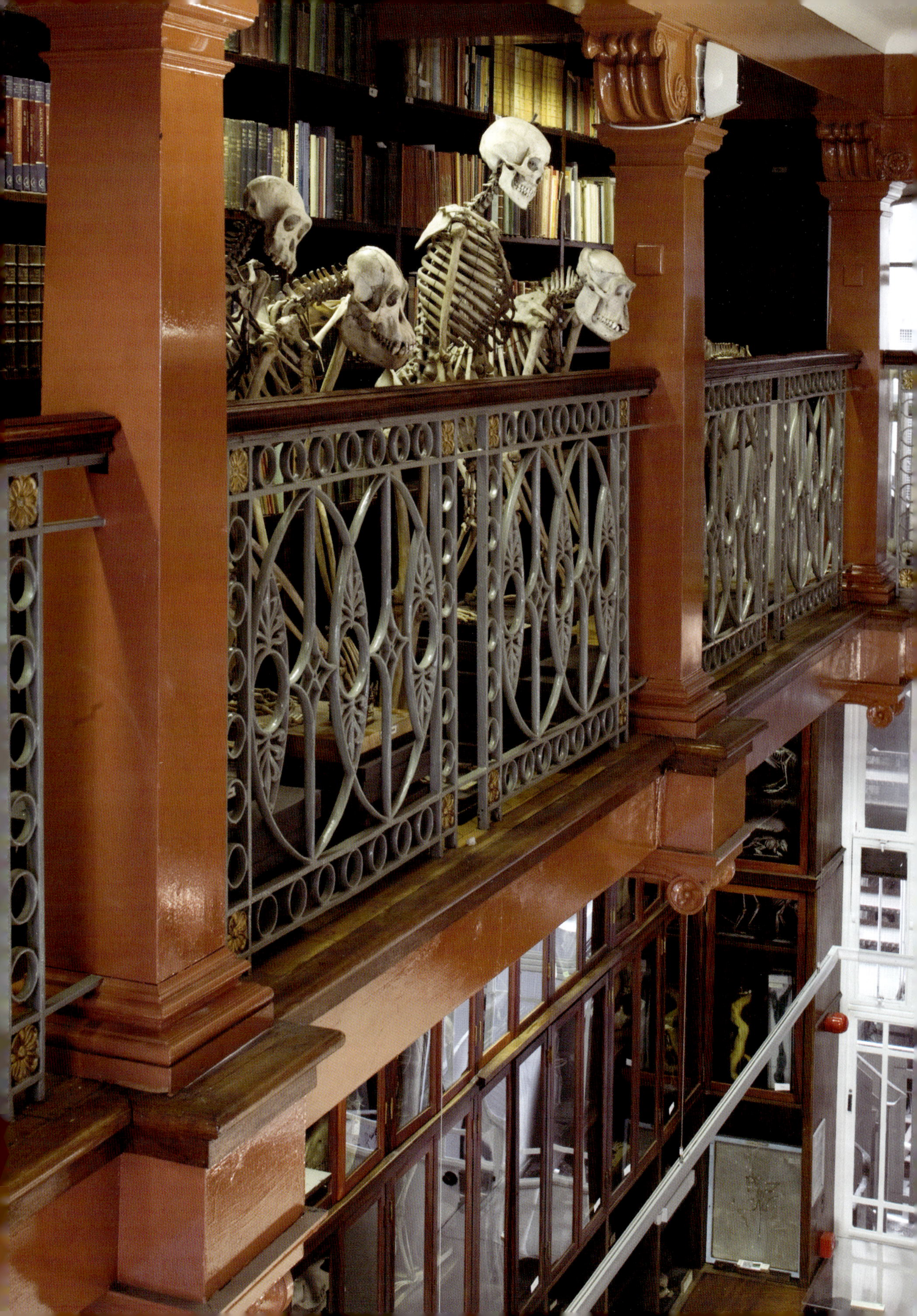

cognitive neuroscience have led to possibilities which could not have been imagined in the wildest dreams of curators just fifty years ago. Our curators are also using 3D-scanning technologies to create new kinds of digital exhibitions, and are experimenting with the very concept of the exhibition, creating single-object exhibitions, virtual exhibitions, and 'pop-up' exhibitions that last for only one hour.

The physical space of each museum is also being transformed. Traditional cases for display are kept, but there is also a new creative and even mischievous approach to displays, as in the Grant Museum, where the skeletons of primates look down quizzically at the visitors or swing lazily from the balcony's handrails. Artists are invited in for residences to use the museums as their studio, and the museums are used for social events as well as academic ones. The museum is also taken physically out of UCL in the form of loan-boxes containing artefacts, which are used in schools to help children understand topics as diverse as citizenship or zoology.

Behind all of this experimentation is an abiding commitment to reach out to new and traditional audiences in order to explore objects in different ways – and thereby to understand both history and our world in a better way.

UCL has in its museums and collections many objects with fascinating stories behind them. It is also fortunate in having extraordinary curators who care deeply about conserving, preserving and exhibiting the collections. And, faithful to the radical tradition of UCL, they are transforming the way the university engages with the public, involving communities in formulating the research questions that will drive forward our major enquiries. Public engagement is about dialogue, about listening as well as speaking. It is also, crucially, about more than expanding and sharing knowledge; it is about shifting and sharing the authority that comes with knowledge. Francis Bacon (1561–1626) famously said that 'Knowledge is power', and public engagement as practised by UCL's curators, scientists and scholars helps to share and democratize the power of knowledge.

Our curators are radical also in encouraging more and more people from more and more backgrounds to listen to their curatorial stories but also, just as importantly, to create

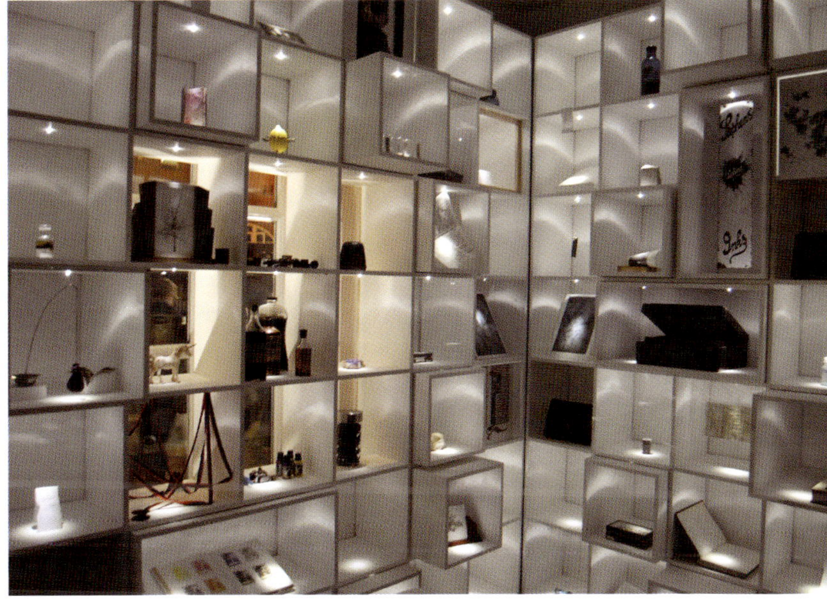

Bringing together art and science, objects and people in the heart of UCL, the 2010 temporary exhibition Ink *in the North Lodge.*

and share their own stories. The museum is no longer a place of single authority, but rather one where many views and interpretations meet and interact.

Francis Bacon also wrote that 'Wonder is the seed of knowledge'. I said above that museums are places of the past and of history; they are also places of looking and thinking today, in the present. They are also places of surprise, places where we discover things that suddenly, unexpectedly, speak to us and help us to make connections. These connections are with other ideas, with memories from the past, with things learned, thought, felt, loved and sometimes lost. This volume is a cornucopia of responses by UCL colleagues to individual objects held in our many museums. Each one is different; each one is personal; each one illuminates an object. Each one is informed by knowledge, yet each one springs out of wonder, love and gratitude. As you read each essay and look at the object that inspired it, you will, we hope, be inspired to come and see the object in person and to wander through our museums, taking the time to look long enough to *see* and to create your own personal memories from the objects.

Primate skeletons on permanent vigil over the Grant Museum.

Contents

p24

p14

p72

p56

p30

p48

p54

p18

p90

p26

p44

p36

p42

p68

p38

p78

p46

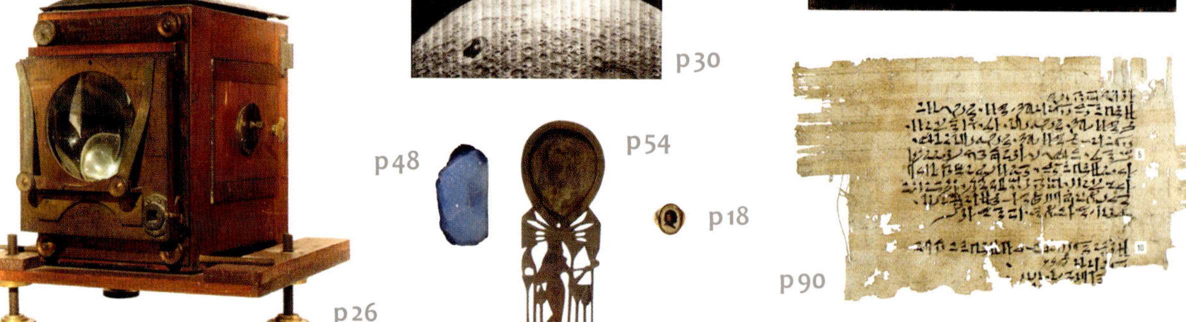

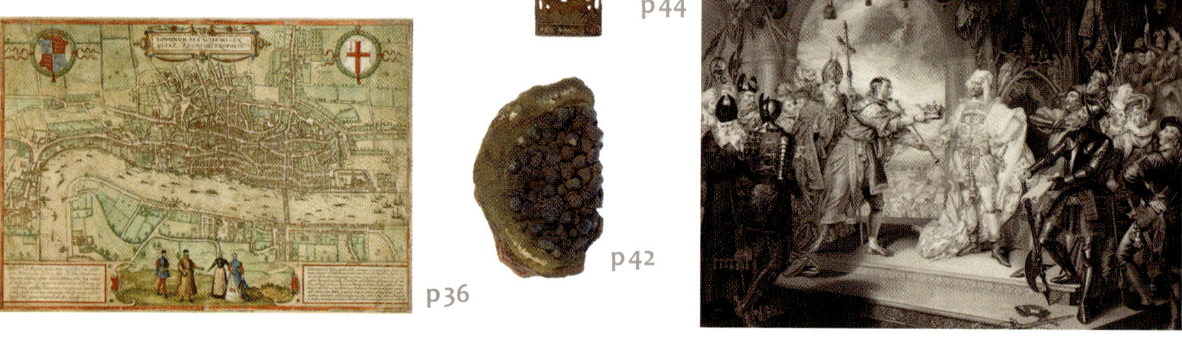

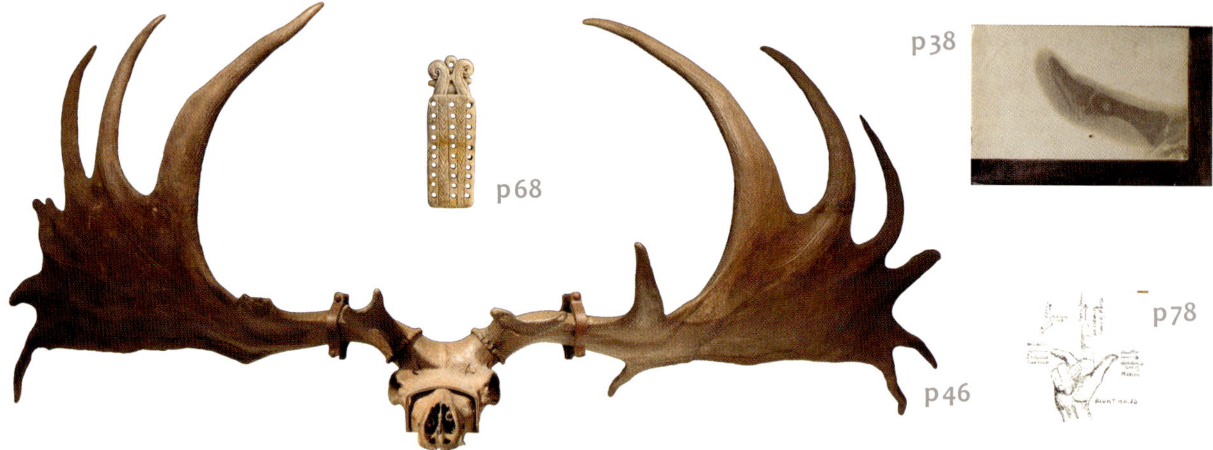

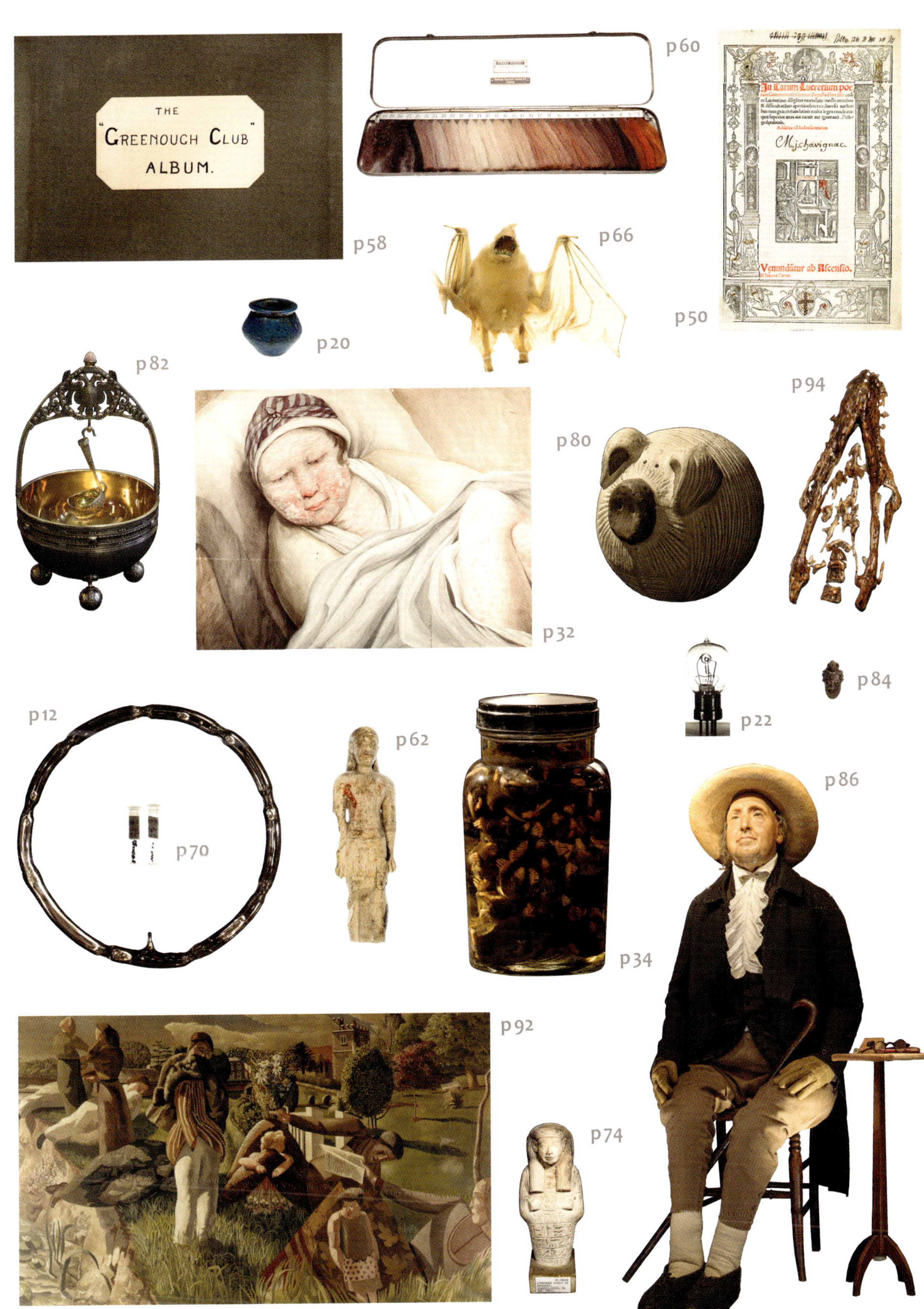

p60
p58
p66
p50
p20
p82
p94
p80
p32
p12
p84
p22
p62
p86
p70
p34
p92
p74
THE "GREENOUGH CLUB" ALBUM.

The Collie Tube

Professor Andrea Sella & Professor Alwyn Davies FRS

Department of Chemistry

S EVERAL YEARS AGO, the Department of Chemistry's draughtsman, John Cresswell, drew our attention to an odd piece of glassware that had languished among a collection of artefacts and odds and ends. The object is a glass tube lovingly shaped into a ring, with constrictions set at intervals along its circumference. The tube, which is sealed off from the atmosphere, contains several millilitres of mercury, but appears otherwise empty. Departmental legend has it that the device was made by the chemist Norman Collie (1859–1942), the fourth head of department who, amongst his many talents, was no slouch as a glass-blower. The astonishing property of this device is revealed when it is shaken: the sloshing of the mercury is accompanied by a delicate crinkling sound, and flashes of greenish and pinkish light from inside the tube. Closer examination shows that the flashes start from the trailing edge of the mercury as it runs jerkily across the surface of the glass. This behaviour was so unexpected that for several days members of the department could be observed creeping into the windowless second-floor toilets and turning out the lights, ostensibly 'to carry out experiments'. A casual conversation with the physics glass-blower, Brian Humm, turned up a second, rather smaller, tube, which he had made as an apprentice. More scuttling about in the dark revealed that this tube only gave the greenish light, corresponding to the visible emission of excited mercury vapour. The pink flashes from the older tube, on the other hand, are due to neon. Collie was not one of the discoverers of neon (although he is sometimes credited with the invention of the neon lamp beloved of advertisers everywhere), but the neon in the tube must date from shortly after the heroic discovery of the element in 1898, by William Ramsay (1852–1916) and Morris Travers (1872–1971).

How does it work? Flashes of light from mercury were first mentioned by the French astronomer Picard in 1676, and further studied by Daniel Bernoulli in 1700. Collie himself described the effect in a paper entitled 'Note on a curious property of Neon' in 1909.[1] More recently, Professor Seth Putterman and his colleagues at the University of California, Los Angeles, reinvestigated the phenomenon.[2] Their experiments show that as the mercury flows across the glass surface, electrostatic charge is produced and separated, causing an attraction of the mercury to the glass. When a spark is eventually produced, the mercury flows freely for a moment, accompanied by the production of a burst of soft x-rays, and the transient excitation of the surrounding molecules, either gas phase mercury or the neon; hence the flashes accompanying the jerky motion. But the origin of the charge separation remains a mystery – contrary to popular belief the electrostatics of bad hair days or the shocks you get from a doorknob after walking on a carpet, remain quite poorly understood. As the great UCL geneticist J. B. S. Haldane once said, 'The universe is not queerer than *we* suppose. It is queerer than we *can* suppose.'

---

1. N. J. Collie,'Note on a curious property of Neon', *Proceedings of the Royal Society of London* (82, 1909), p.378.
2. R. Budakian, K. Weninger, R. A. Hiller, S. J. Putterman, 'Picosecond discharges and stick–slip friction at a moving meniscus of mercury on glass', *Nature* (391, 1998), p.266.

Collie tube
1900
Norman Collie (1859–1942)
Glass and mercury, diameter approx. 28 cm
UCL Chemistry Department Collections

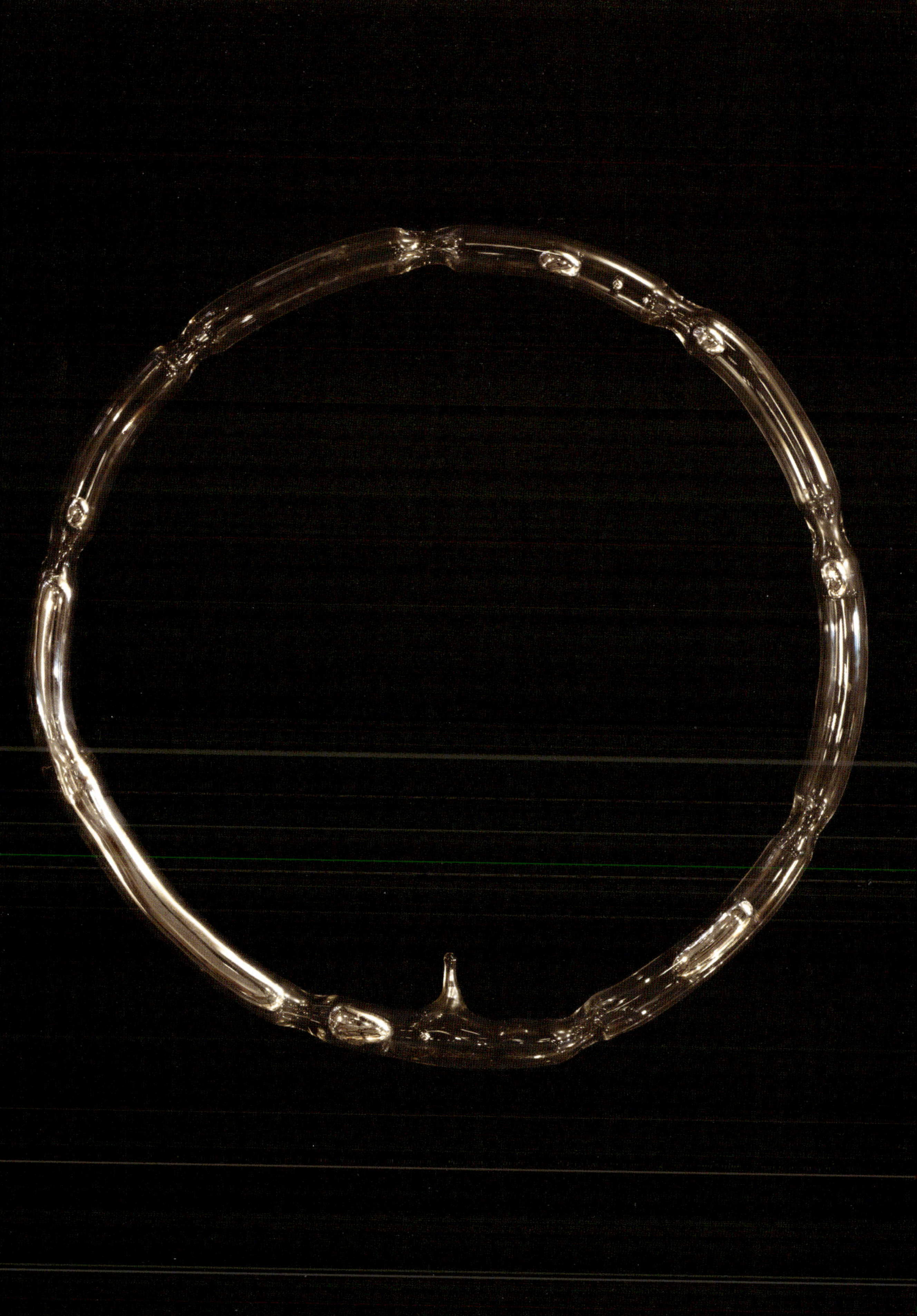

Cinnabar:
'A beautifully red stoney substance that breaks into shining bright, angular pieces'[1]

Dr Ruth Siddall

Department of Earth Sciences

CINNABAR IS A naturally occurring mineral with the composition mercury(II) sulfide. It is a beautiful colour – a deep, slightly pink, red, the same shade as a ruby. However unlike rubies, cinnabar is soft – it can be scratched with a fingernail – which renders it useless as a gem stone. Cinnabar is not particularly common but it is abundant in many of the localities in which it occurs. It is most commonly encountered in two main geological environments: in association with the mineralization of gold veins and as a product of volcanically driven geothermal activity; therefore it is found in the vicinity of certain hot springs. More often than not, cinnabar occurs as earthy masses and is often associated with native mercury, quicksilver, which bleeds out of the host rock. In some instances, as in the image here, cinnabar forms relatively large, well-developed, transparent crystals. This example is in association with the mineral dolomite and comes from Hunan Province in China.

Despite cinnabar's inadequacy as a gem, its beautiful colour remains desirable. The softness of this mineral means that it is easily pulverized, and remarkably, it retains its intense colour when finely ground. Therefore it has been prized as a pigment which, when mixed with a medium, produces an intense red paint. Notwithstanding the risks of mercury poisoning, cinnabar-based paints have been valued by almost every global culture.

Earliest uses of cinnabar as a pigment are in the Neolithic wall-paintings of Çatalhöyük in central Anatolia. In China this pigment is known to have been in use from the second millenium BC, identified on oracle bones. To the Romans, cinnabar obtained from the mines of Almaden in Spain was one of the most valuable commodities. Prices were capped by law at 70 sesterces a pound to stop costs spiraling out of all proportion. At over ten times the price for a high-quality red ochre, cinnabar was a symbol of wealth and taste. This is the colour that gave us the phrase 'red letter day' from the Romans' use of cinnabar to write important dates on calendars and inscriptions.

During the European Renaissance, the value of cinnabar was second only to that of blue pigment ultramarine. It was used to paint the robes of cardinals and those of the Madonna. The technique of manufacturing this compound from mercury and sulphur had been developed and this synthetic form of mercury(II) sulfide goes by the name of 'vermillion'. The natural and synthetic forms are difficult to differentiate, but both produce a strong, dense and beautiful red.

However cinnabar and vermillion are not the perfect reds. Some forms of cinnabar turn black when exposed to light. The exact mechanisms for this remain obscure; however it is likely to be related to trace amounts of chlorine in the crystal structure. Nevertheless this unfortunate property was described by the Roman architect Vitruvius in the first century BC and it is a little known fact that the vaults of many of Europe's art galleries contain portraits of cardinals, no longer fit for display, because their beautiful red robes have now turned an unsightly black.

---

1.   H. Boerhaave, *Elementa Chemiae* (Leiden, 1732).

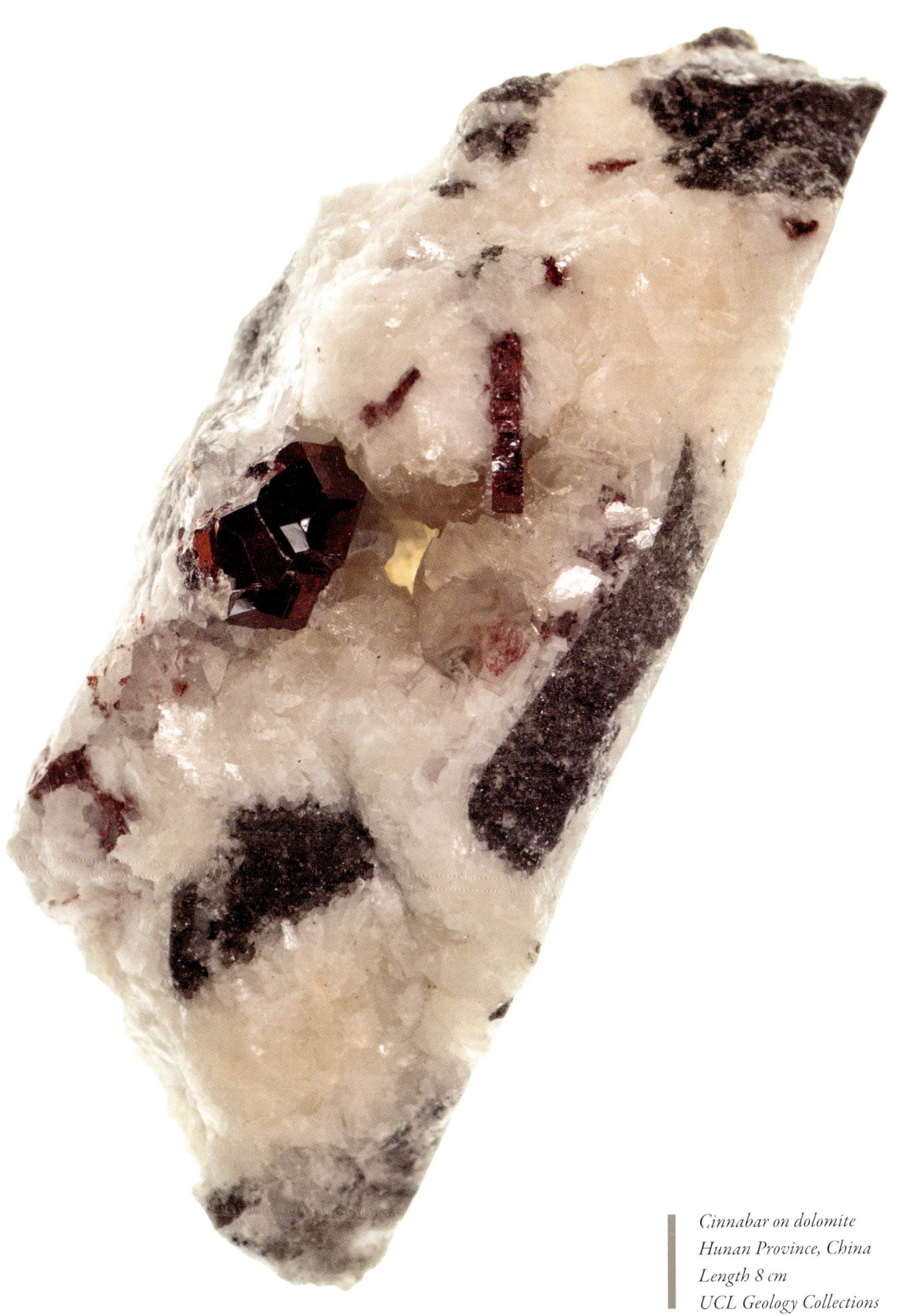

Cinnabar on dolomite
Hunan Province, China
Length 8 cm
UCL Geology Collections

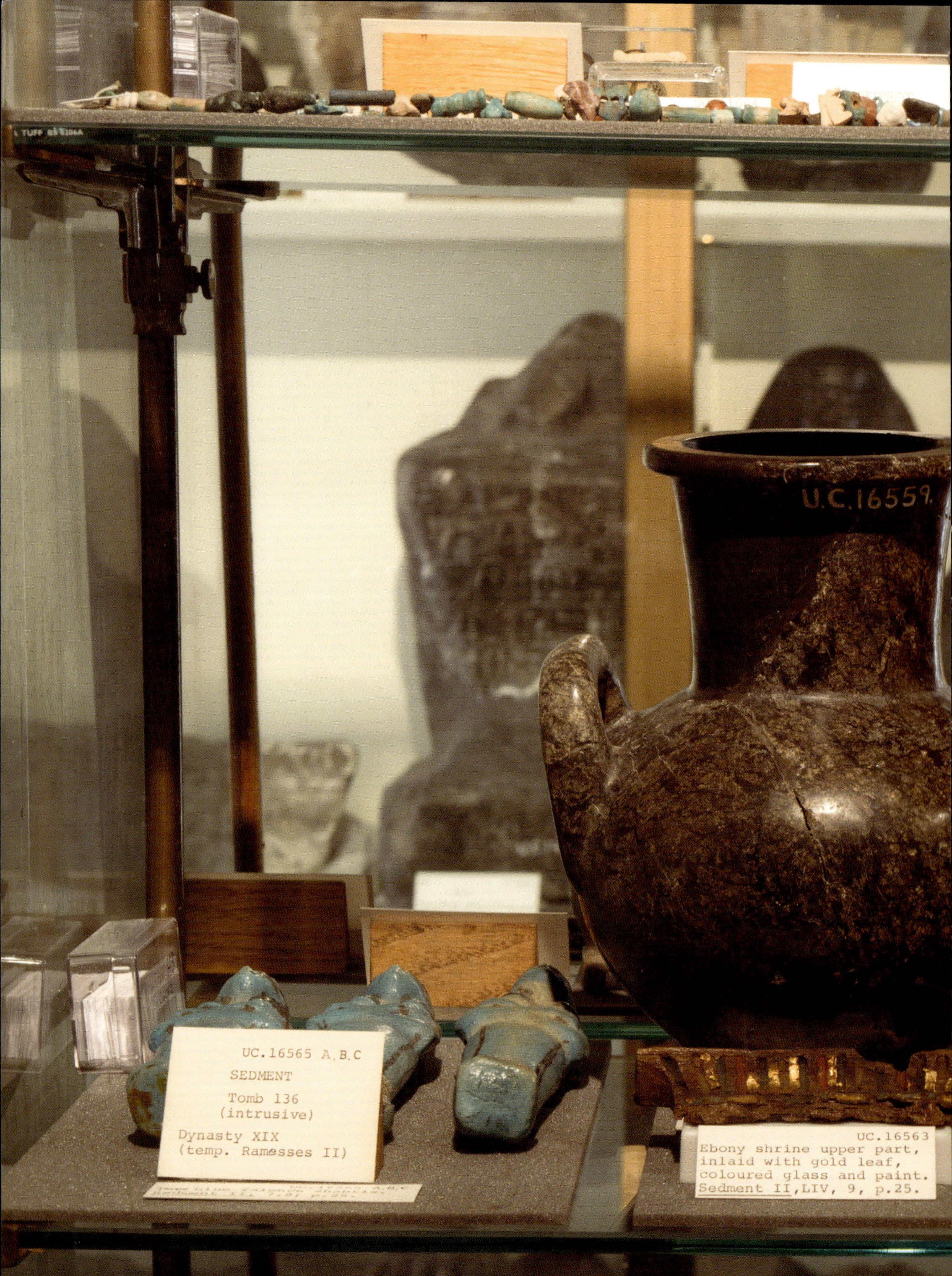

U.C.16559.

UC.16565 A,B,C
SEDMENT
Tomb 136
(intrusive)
Dynasty XIX
(temp. Ramesses II)

UC.16563
Ebony shrine upper part,
inlaid with gold leaf,
coloured glass and paint.
Sedment II, LIV, 9, p.25.

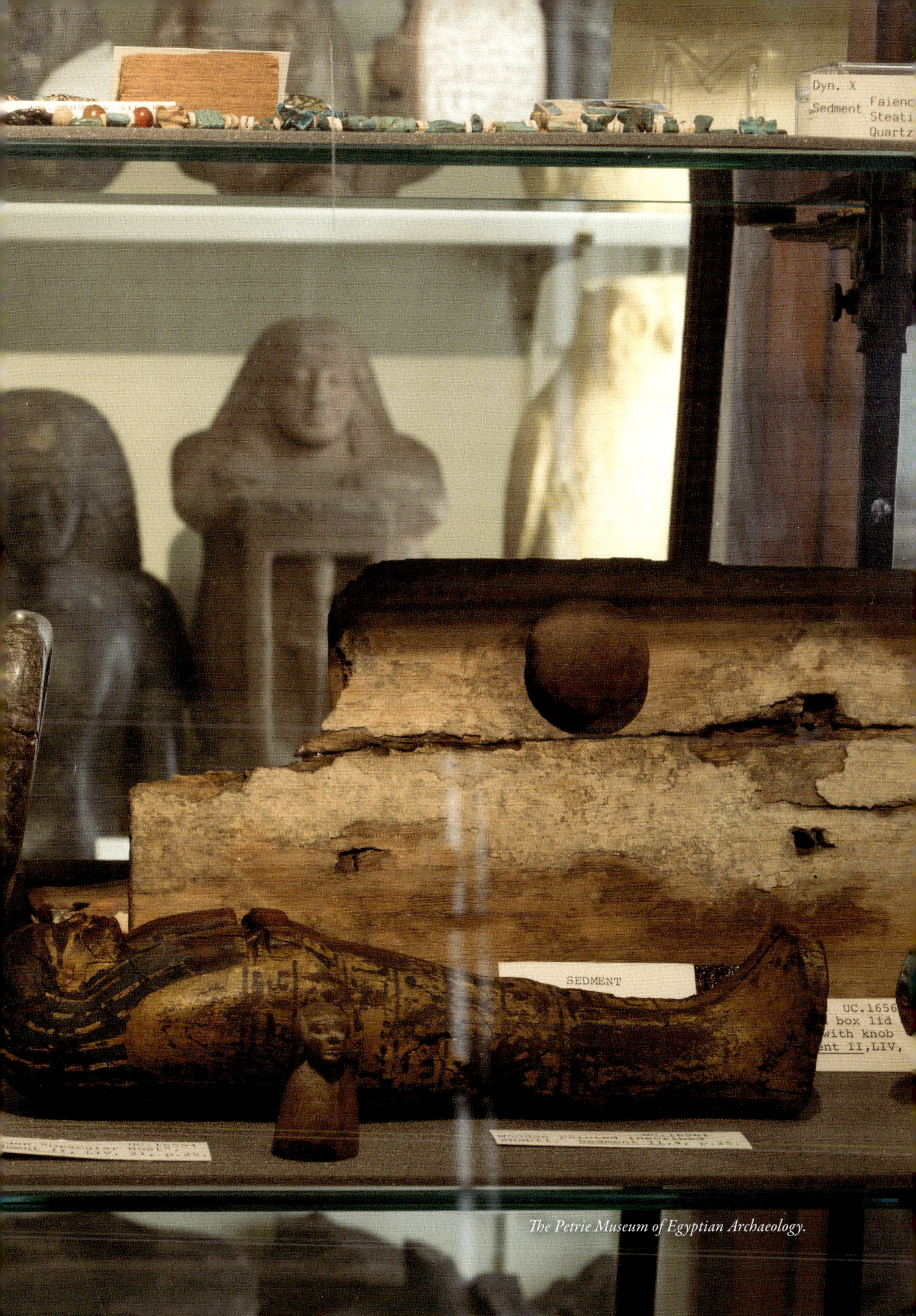

The Petrie Museum of Egyptian Archaeology.

Lord of the (Bentham) Rings

Professor Philip Schofield

Faculty of Laws

T HE DARK LORD Sauron, in J. R. R. Tolkien's *The Lord of the Rings*, gave magical rings to nine mortal men, seven dwarf-lords, and three elven-kings, while he kept the most powerful One Ring for himself. He had the misfortune to lose his ring, along with his finger, in battle, and the ring remained lost until discovered by Gollum. Sauron wanted it back.

Jeremy Bentham gave – or rather left in his will – twenty-seven rings. These mourning rings were distributed after his death on 6 July 1832 to relatives, friends and employees. When I started my career at the Bentham Project in 1984, I did not expect to become, like Sauron, a collector of rings, but I have played a part in the acquisition of three Bentham rings for UCL, to add to a ring that the university had acquired in the 1930s. This ring was donated by the descendants of John Bowring (1792–1872), Bentham's literary executor, who

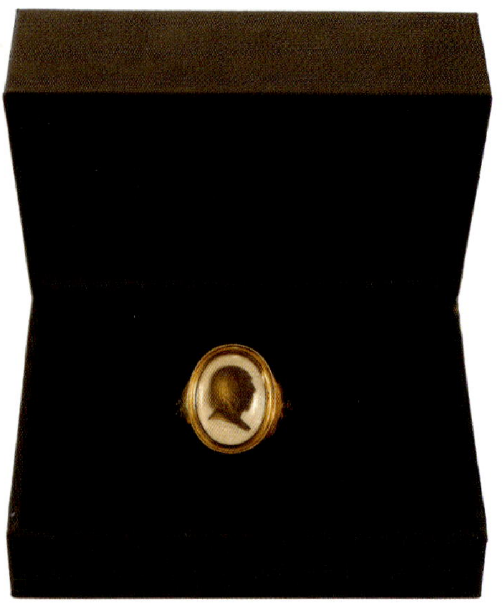

had superintended the production of the 19th-century edition of *The Works of Jeremy Bentham.* The inadequacies of the Bowring edition led UCL to establish the Bentham Committee in 1959 in order to superintend the new authoritative *Collected Works of Jeremy Bentham* (twenty-nine volumes published, another forty to go!). The ring, which is displayed with Bentham's auto-icon, is made from gold, with a black and gold painted silhouette and a locket compartment containing Bentham's hair. It is not, however, inscribed to a particular individual, unlike the three that I have helped acquire – it was possibly surplus to requirements, and so retained by Bowring.

I bought the first Bentham ring at auction in 1997. This ring had belonged to William Tait (1793–1864), the Edinburgh publisher and bookseller, who had published the Bowring edition.

The second ring, which had belonged to the Belgian politician Jean-Sylvain van de Meyer (1802–74), had been acquired by Denis Roy Bentham, not a descendant of Bentham, but a postal worker from Loughborough and an astute and determined collector of Benthamiana. Following a visit to Mr Bentham in 1999, I wrote to ask him to consider leaving his collection to UCL. He died later that year, whereupon his sister, having read my letter, kindly gave UCL around ninety original Bentham manuscripts – and the Meyer ring.

The third ring, and perhaps the most 'precious' as Gollum would have said, had belonged to the philosopher John Stuart Mill (1806–73). The ring was purchased by Mr Michael Phillips, an alumnus of the Faculty of Laws, in a jeweller's in New Orleans. Mr Phillips visited the Bentham Project in 2007 and, without fuss or ceremony, presented me with the ring. This ring is special because it represents a physical link between the two greatest utilitarian philosophers – a handing down of intellectual leadership.

A fourth ring is in the possession of descendants of one of Bentham's servants. That means that there are twenty-three rings unaccounted for. If you have one, please get in touch. I promise I won't send Gollum round to snatch it.

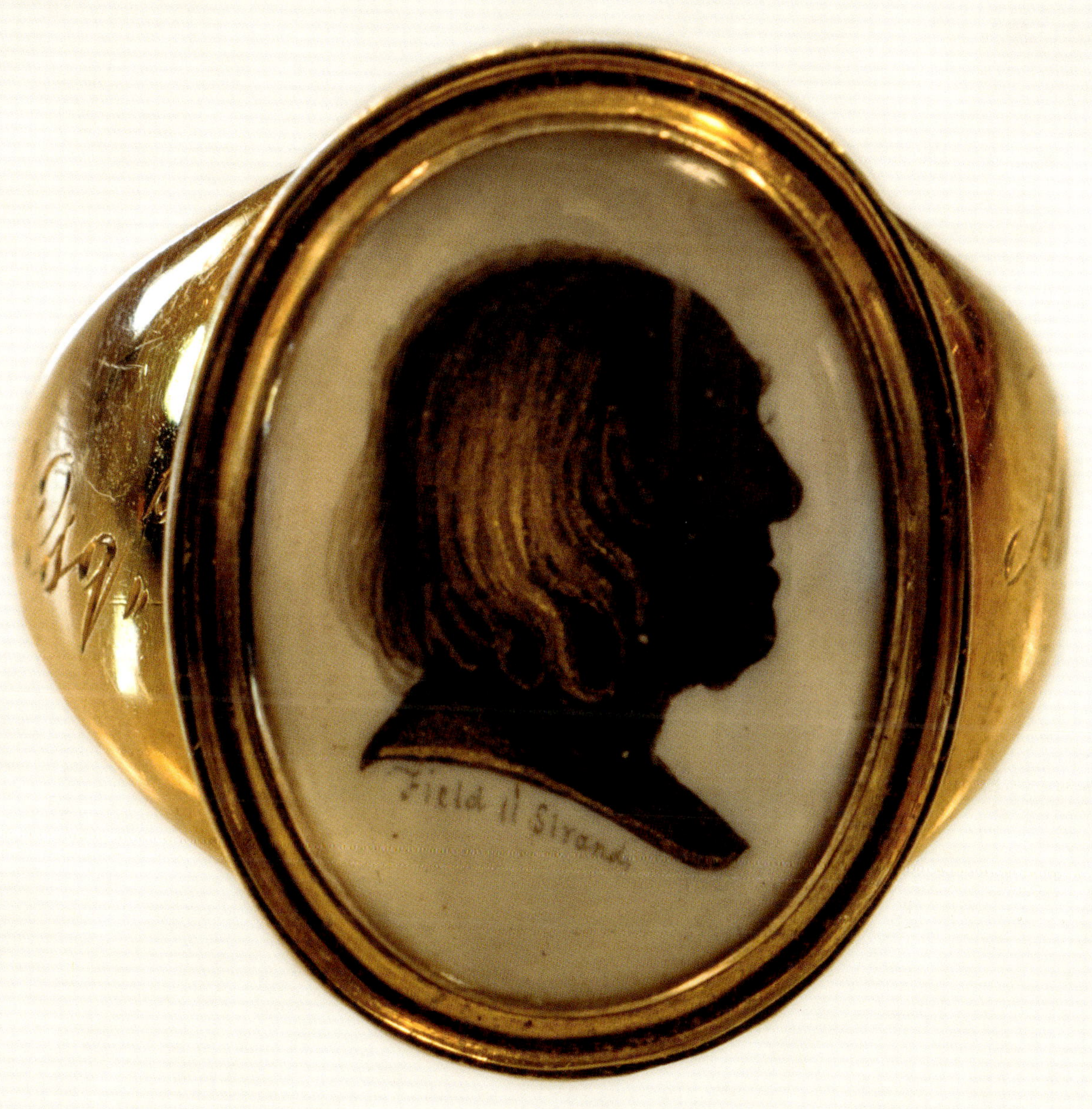

Field il Sirand

Tangled up in blue?

Dr Anne Lanceley

Institute for Women's Health

I WAS DRAWN BY the accidental beauty that this object generated, placed next to pale alabaster and some honey-coloured beads in the Petrie Museum. Once I had seen the tiny pot with its rich dark blue glaze it kept slipping back into my mind to represent wholeness and in some small way a victory of intimacy over sterility and humanity over inhumanity. This very small everyday object, which would have been held close for the application of kohl to an Egyptian eye, or perhaps scented oils to a troubled brow, had well and truly chosen me. Its thick dark glaze evoked the lapis lazuli I first encountered when travelling in central Asia as a teenager. Spidery labels and marks on the pot reveal it was from excavations directed by Flinders Petrie himself in 1898–99. It was found in a 2200–2000 BC burial on the west side of the Nile, sixty kilometres or so north of Luxor. Attracted to the romance of this archaeology, I grew slightly nervous as I gazed at the object, as if it had begun to show itself for a conscious thing, watching my state with enigmatic grace.

Over the several months this book has been in the making I have returned to the kohl pot in my mind and been surprised by the direction of my thoughts in relation to it. These have centred on my choice of work with women with cancer; how I feel caring for ill women; and how I understand and attempt to deal with my own feelings in order to continue my nursing work and research. My desire to nurse and improve the treatment and care of women with gynaecological cancer has – it is claimed by psychodynamic thinkers – unconscious determinants such as my own inner and unresolved conflicts and need for repair. The pot is so satisfying to me I think because symbolically it represents repair. The value and pleasure of the object is the result of the craftsman successfully overcoming his internal conflicts and anxieties in order to make it. The object is restorative to me in its simplicity and completeness, and it provides comfort. I instinctively identify with its smallness and I feel hopeful when I look at it, for despite its smallness it has survived. The object has been so beautifully crafted it provides me with a creative, life-affirming counterbalance to loss, which is an inherent part of my work. Beholding the pot is therefore an act of reclamation.

The pot punches an emotional weight far above its size and is an utter delight to my eye.

Blue faience keeled kohl pot
about 2200 BC
Faience, diameter 4.9 cm
UCL Petrie Museum of Egyptian Archaeology
LDUCE-UC18926

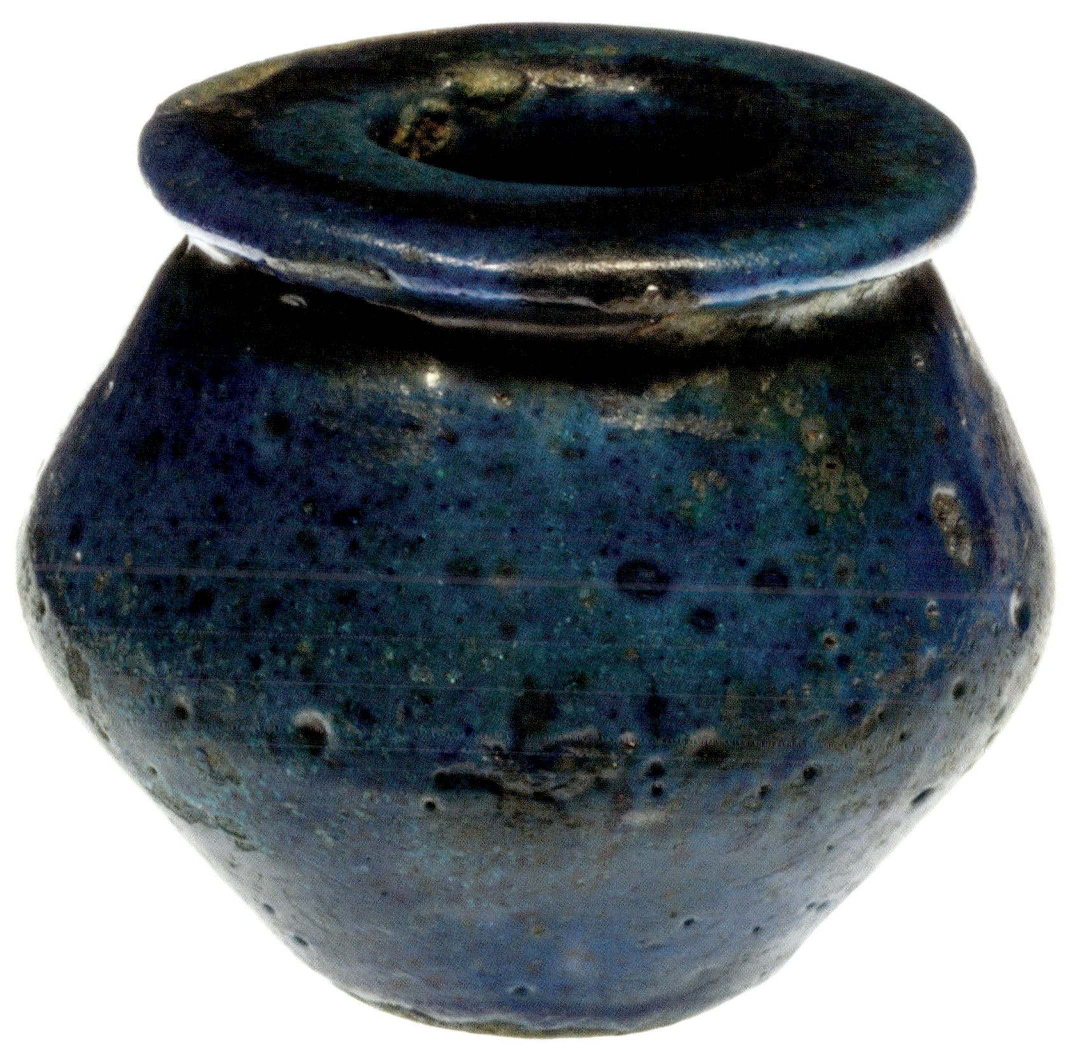

Valve: an electrical romance

Professor Anthony Finkelstein

Dean of the Faculty of Engineering

I REMEMBER, AS A CHILD, lying in my bed with the curtains drawn and the lights off, in the glow of an old valve radio that I had salvaged from the loft. It had a distinctive smell, when switched on, of dust burning on the valves. Now when I smell this I am transported, as Proust with his madeleine, to my childhood.

The illuminated dial had a romance of its own with, even then, long-dead broadcasting stations – Hilversum, Droitwich, and so on. I would tune to short wave and find obscure radio stations, Soviet Radio Kiev, with the latest tractor production statistics. After the news, Vladimir Pozner, in flawless American-accented English, told us the truth we had not been allowed to know. But my favourites as I drifted towards sleep were the numbers stations: just above the static, at the far end of the dial, punctuated sequences of numbers in distant female voices, often introduced by tones or small fragments of simple music – the 'Lincolnshire Poacher' – identifying the stations. Messages to spies, perhaps, beamed to my small, suburban bedroom.

Hearing my parents coming upstairs I would hurriedly switch off, turning the mechanical volume knob until it gave a loud click. Though the dial would go dark, the valves, visible through the vents on the top of the radio would still glow, slowly cooling. I worried that the dull orange, reflected off the ceiling, would give away my, supposedly secret, listening.

I built my own radio, of course, with components from Radio Spares. Not using valves it lacked, how shall I put it, sensuality. Soldering was fun, though. I listened to it crouched under the sink where I had, to my mother's disapproval, scraped the paint off the copper pipes to get a decent earth. With earphones I could catch Radio 1, listening to Mud and The Sweet, until the discomfort of my position led me to abandon it and go to bed. I can't quite remember what happened to the old radio. I think a valve finally blew and lack of spares consigned it to the bin.

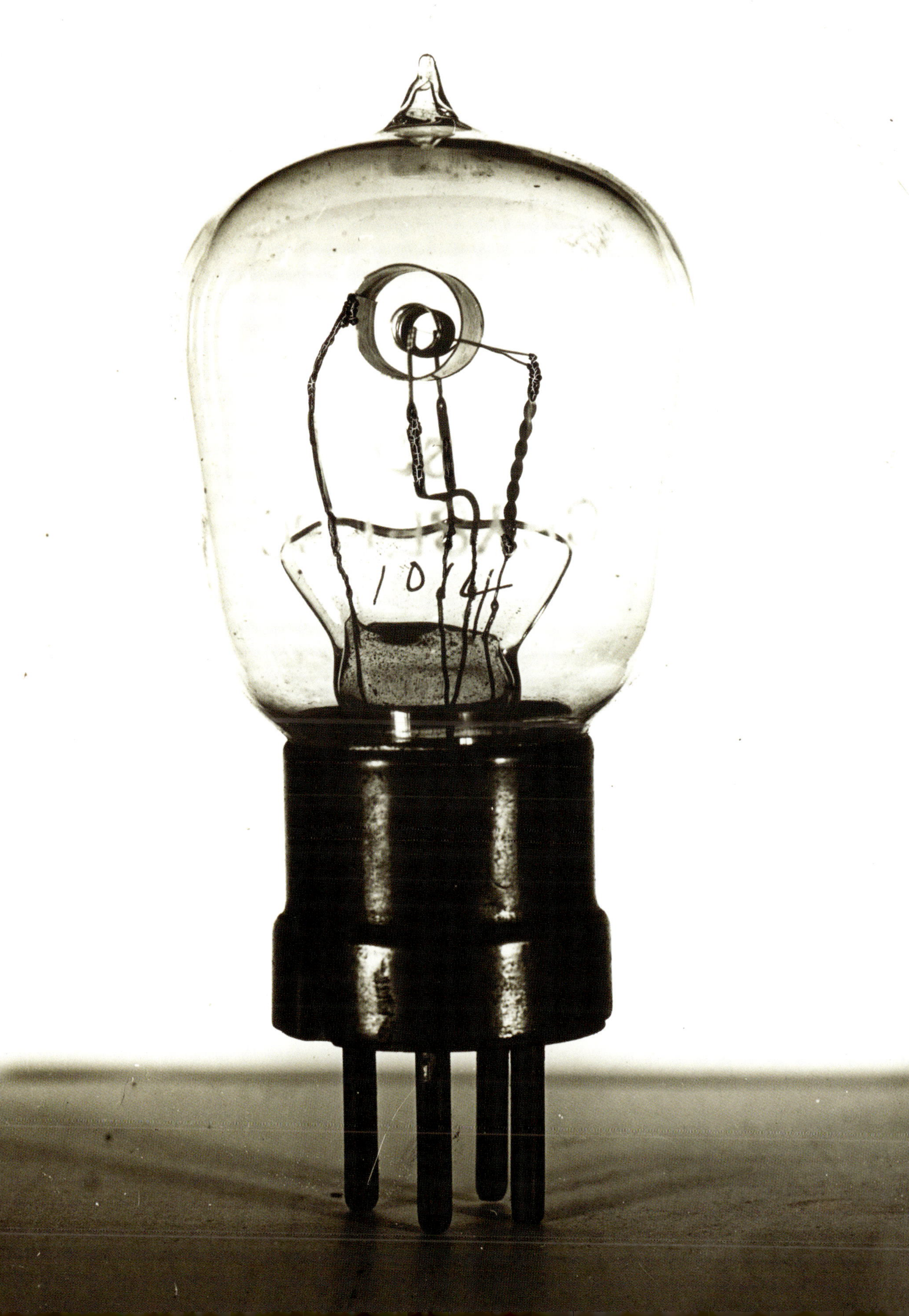
1024

Black Life in London

Dr Caroline Bressey

Department of Geography & Director of the Equiano Centre

I LOOK IN ON THE MAN looking out, sitting in the chair in his shirtsleeves in a studio, somewhere in the Slade School of Fine Art in 1919. Although they seem relaxed, his large hands with their long elegant fingers hang down rather awkwardly from the ends of his shirtsleeves. He looks out beyond the gaze of those picturing his presence. I look into his face, seeking to read his mind. Uncovering the lives of black working men and women in inter-war London beyond the lines of census statistics is hard. I try to unpick the thoughts forming behind the stillness of his dark alert eyes.

Perhaps he is thinking about what he will cook for dinner that evening in the lodging-house kitchen, once the students at the Slade have finished copying his likeness. Or perhaps he is not worried about dinner for that is someone else's job. Instead his thoughts turn to his children, and how different their lives are to the young men and women who stand about him in the room – his youngest son fighting to find peace now he has returned from the front; his oldest daughter struggling bringing up a child alone as her husband never came home. The year after the end of the war to end all wars has proved to be violent and bloody. Workers have been on strike, men and women fighting in uprisings, and race riots riding waves of anger and frustration across the British Empire, Europe and the United States. He remembers when he first heard about the first race riots in Britain, the ones in Glasgow back in January. Sitting in The Lamb he and his friends had shaken their heads in disbelief. By August, South Shields, Salford, Hull, London, Liverpool, Newport, Cardiff and Barry had all shown that men and women across the country were happy to turn against their African, Indian, Chinese, Caribbean and Arab neighbours. No one knew how many people had been injured or if it was all really over yet. Five people had been killed. They said over two hundred had been arrested, but that didn't help anyone feel safe – not when they arrested people trying to defend themselves, just trying to get by. When Charles Wootton was chased into the waters of the Queens Dock in Liverpool the police were there on the banks, but they didn't arrest anyone then. He had been just twenty-four, only two years younger than his son.

In his mind the man straightens his back and lifts his chin a little higher. He is proud but tired now. The studio is quiet and still and safe. But perhaps the riots haven't erupted in London yet. Perhaps sitting for a class is just a better way to pass the time than sitting in the pub, a way to make some extra money while sitting in a chair, in his shirtsleeves, in a studio, somewhere in the Slade in 1919.

Portrait of a Man in His Shirtsleeves
1919. Ivy MacKusick
Oil on canvas, 76.2 x 50.8 cm
UCL Art Museum
LDUCS_5090

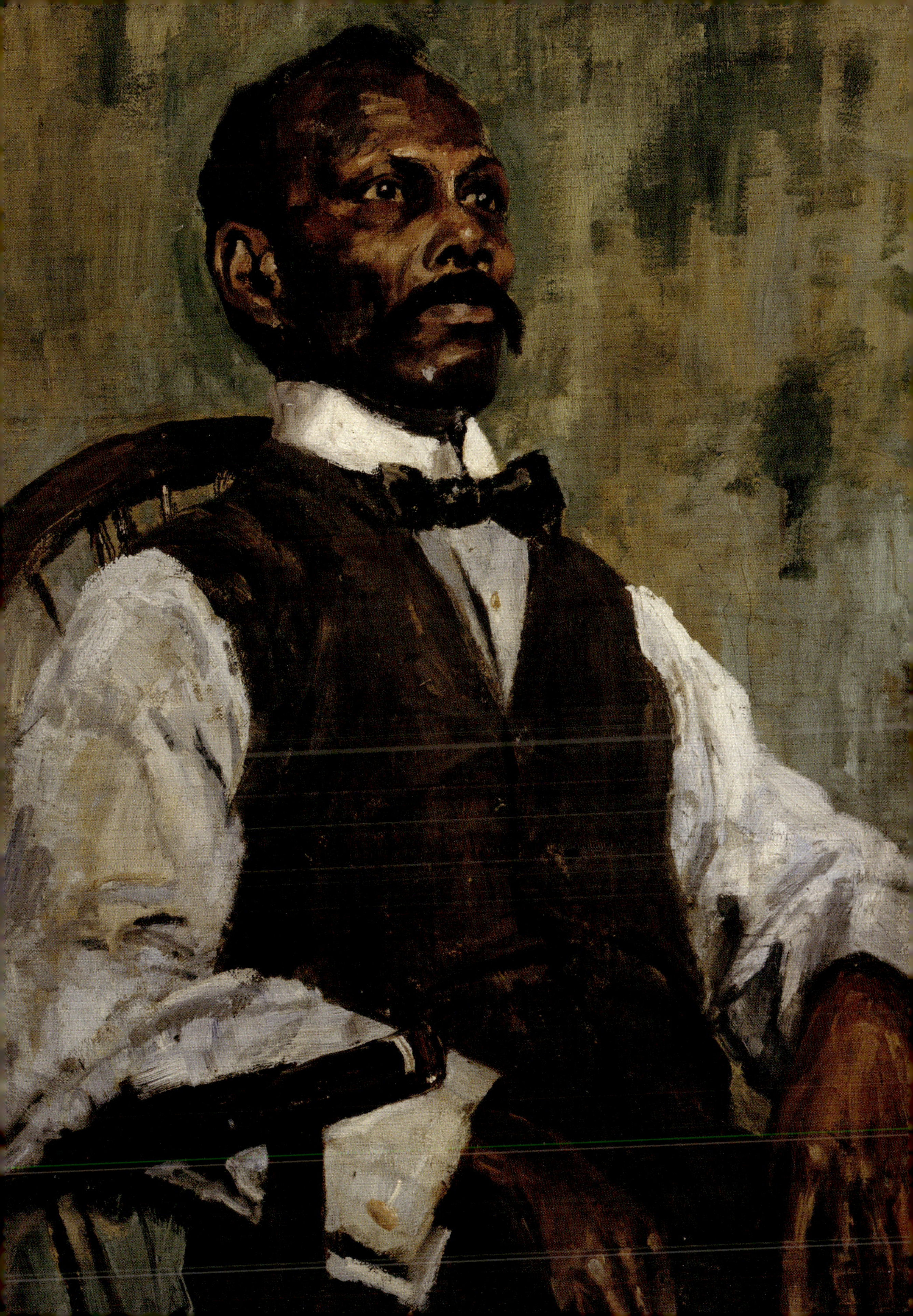

Laterna Scandinavica: Encounters with still and moving images

Dr Claire Thomson

Department of Scandinavian Studies

A STILL SUMMER'S DAY, 1840. The sculptor Bertel Thorvaldsen has been persuaded to sit for a daguerrotype portrait. By a quirk of over-exposure, the relief sculpture beside Thorvaldsen will be obliterated when the image is developed. Meanwhile, every detail of the leaves behind him, motionless in the still air, will be captured for posterity in the daguerrotype's alchemy of silver and mercury. Thorvaldsen is making the Italian *corno* gesture, the first and little fingers extended from a clenched fist, warding off the evil eye.

Winter 1894, Dornach, Switzerland. August Strindberg is working feverishly on a technique that will achieve what the human eye or camera lens cannot: reflect the heavens as they really are, undistorted by the concave shape of the

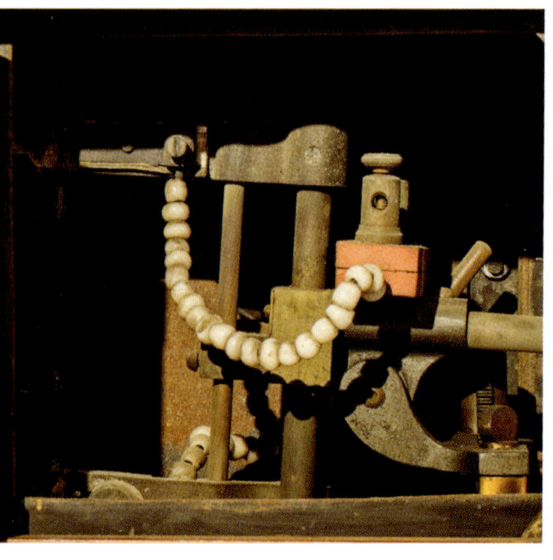

eye or lens. His attempt to capture the sun has resulted in a photographic plate covered in tongues of fire. Now he lets a sheet of photographic paper lie on the window sill, directly exposed to the night sky for forty-five minutes. When this *celestograph* emerges from the developing bath, he will see whorls of galaxies, and a moon misshapen like a bean. He will conclude that this image, free from the demonic interference of the lens, shows the luminaries as they really are: *not round.*

Bernstorff's Palace, Copenhagen, 1899. King Christian IX of Denmark, Tsar Nicholas II, Princess Alexandra, Prince Edward and King George of Greece gather for a commemorative photograph. The photographer, Peter Elfelt (né Petersen), has brought with him a modified cine-camera, the first in Denmark. The royal children spill down the steps and the crowned heads of Europe laugh and tease each other. In a few moments, the lively bodies on the steps will be frozen in a group portrait of dignified formality. But the instant of the photograph has been forever set in motion by the moving image that captures the preceding moments. Thus one of Denmark's first moving images is *a film of a photograph.*

A Danish man disembarks at Christiania harbour, his infant son in his arms. It is 1905. Soon, the title Håkon VII of Norway will be bestowed upon him, and the city, too, will be re-christened: Oslo. The film cameras roll, capturing the moment of a new nation's birth. The film will travel the length of the land, allowing fishermen and farmers to encounter their new king and their urban compatriots, living and moving via the miracle of film projection.

Christmas 1927. The nine-year-old Ingmar Bergman crouches in his nursery closet, assembling the magic lantern he has bartered from his older brother for the price of a few tin soldiers. The apparatus is lit by an internal kerosene lamp and has been modified to carry both glass slides and a 35mm film, hand-cranked on a loop. The boy ignites the lamp, threads the film, and the *laterna magica* throws an image onto the closet's white wall: there is a woman in a meadow, dressed in national costume, and she is *moving*.

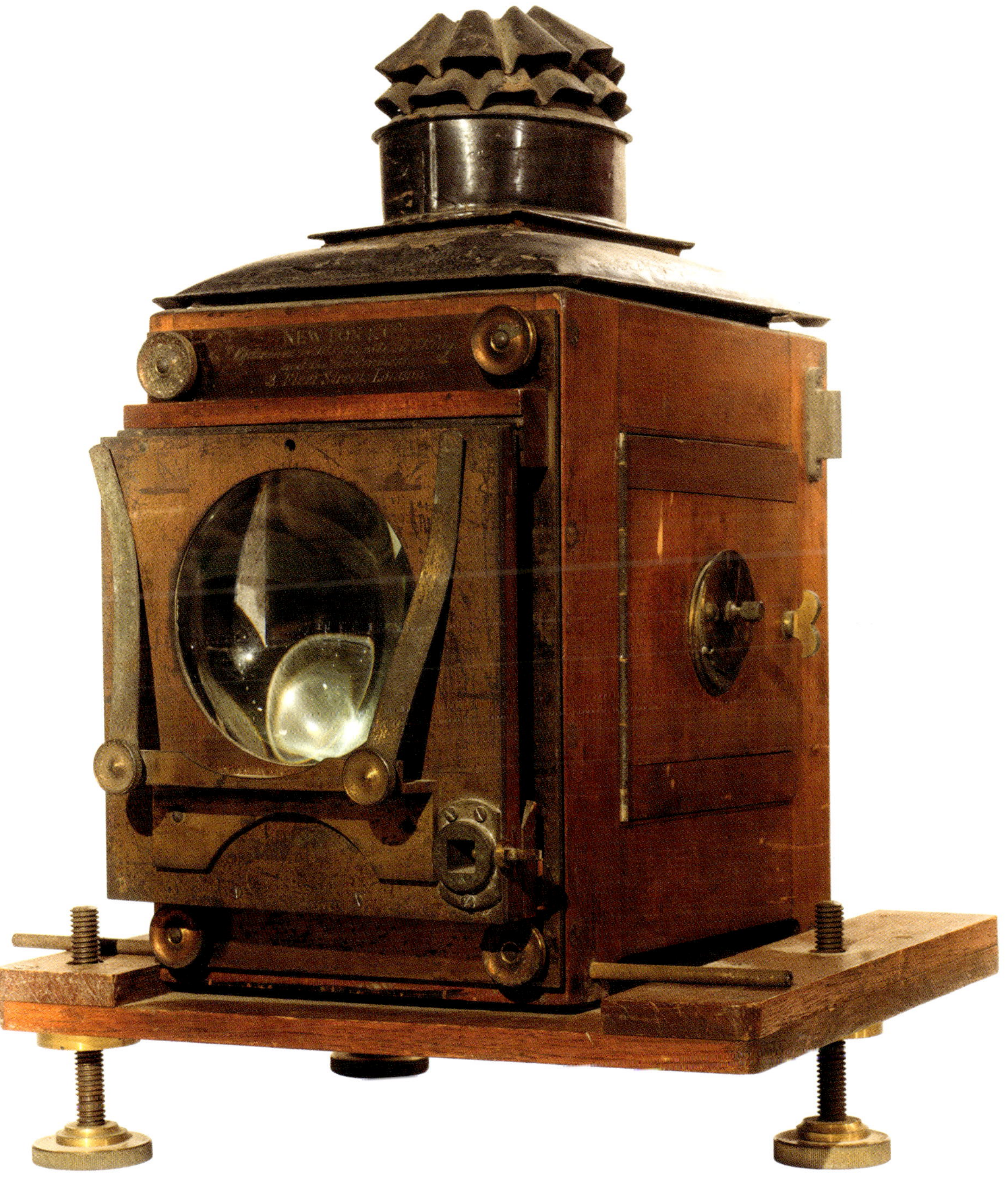

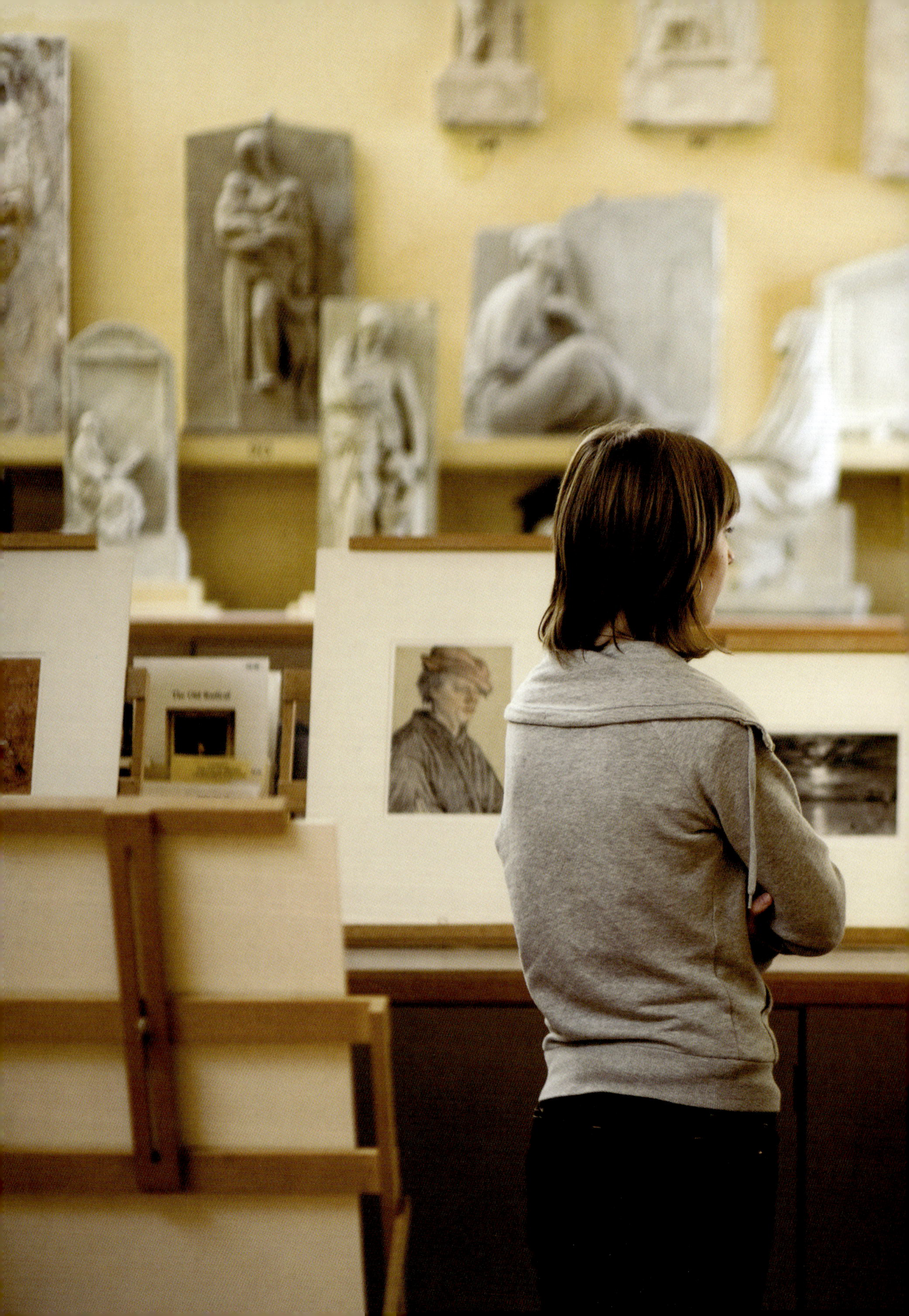

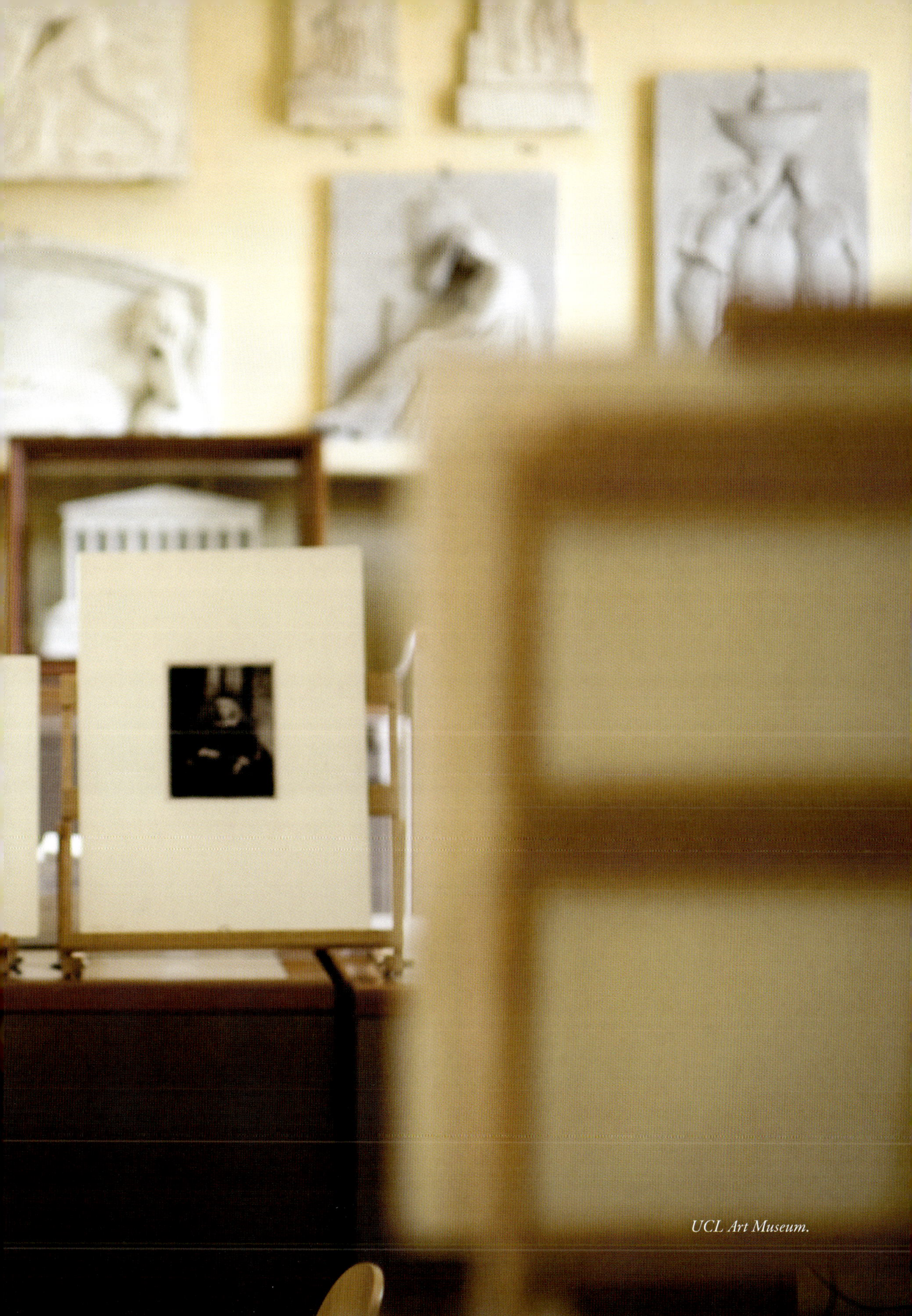

UCL Art Museum.

What are we proud of?

Professor Jonathan Butterworth

Department of Physics & Astronomy

THIS IS A PICTURE of our home planet, Earth, taken from a lunar orbiter.

To me it is about our place in the universe. Depending upon my mood it affects me in two ways.

Firstly, while I am not a religious person, I was brought up as a Christian and I know the value in the 'view from afar' that religion can sometimes provide. It is a sense of perspective and space. Durham Cathedral can do it. This image does it. All the business and busyness of everyday life are in that slim crescent. It may seem overwhelming from down here, but surely we can cope. It doesn't diminish the value of the things we care about, that we love. In fact quite the reverse – see how precious they are! But it gives a breathing space and a sense of wonder.

In a different mood, there is a sense of approaching. Imagine a visitor coming to Earth, over that horizon. How would we react? What do we have to show for ourselves, a struggling species on an island in space?

I carry out research at the multinational CERN laboratory, in Geneva, Switzerland. The Large Hadron Collider is a worldwide effort, trying to extend our knowledge of the fundamental workings of the universe. Thousands of people from all around the world, funded by contributions from billions of people, work together just to learn more about the confusing, wonderful place we find ourselves in. The tunnel the beams travel in is 27 km long. The experiments weigh thousands of tonnes and produce about 15 million gigabytes of data per year. In 2012 we discovered a new particle which may be responsible for the mass, the substance, of all fundamental particles.

The project is a sign of our desire to understand, a desire manifest across many amazing areas of research, of which particle physics is just an example. But the scale of this project is unique, and the vision and international cooperation required to realise it are stunning. I would take the alien visitor to see it.

As a planet, as a species, we have terrible problems. Tragedy, cruelty and stupidity are common and we could and should do much better. But while we can come together to do something like this, we have something at least to be proud of.

And reason to hope.

Lunar Orbiter 1 image of the Earth from the Moon
Photographic print
UCL Regional Planetary Image Facility

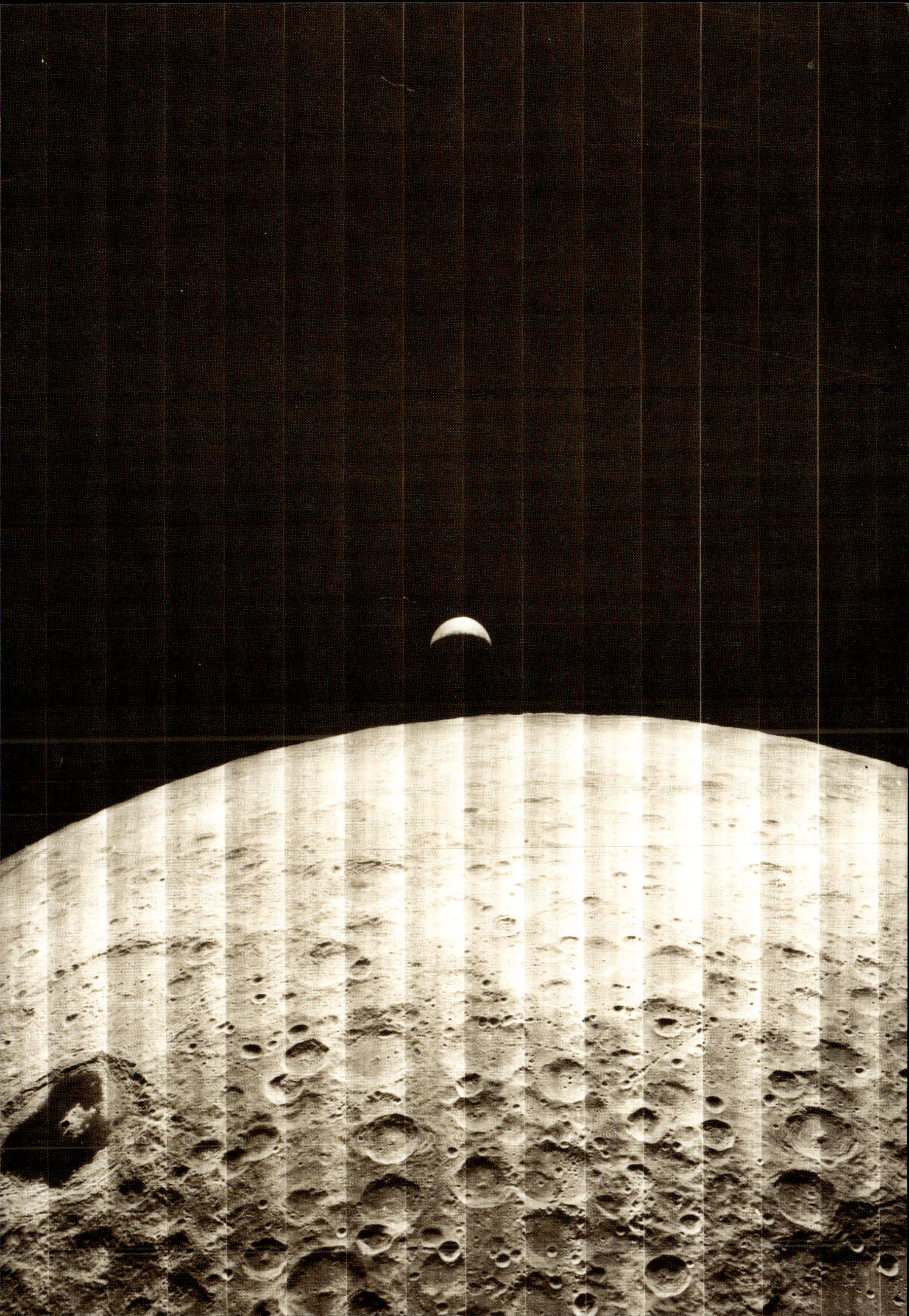

Facing Skin Disease:
a dermatological watercolour
by Robert Carswell

Dr Mechthild Fend

Department of History of Art

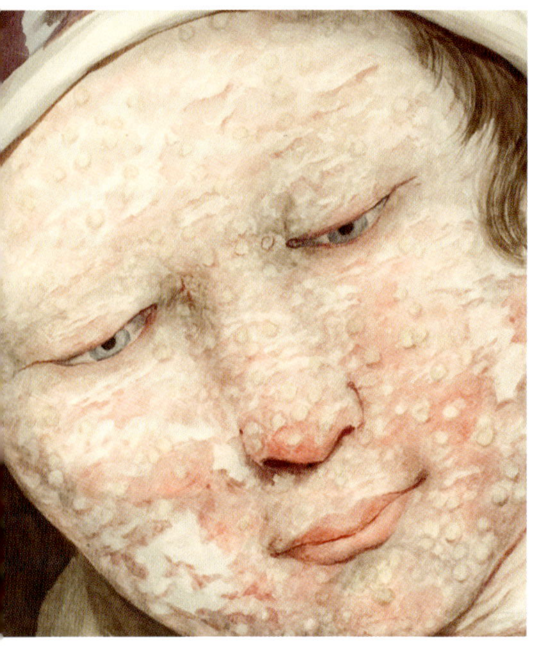

A WATERCOLOUR, DOMINATED by whites and shades of grey, a landscape of monochrome folds, flesh and cushions in which a colourful head stands out. Framed by a scarf with a lively pattern of purple stripes, the face displays white spots on reddish skin. Eyes show through swollen lids without addressing the viewer. Yet, it is hard not to engage with the portrayed, not to feel empathy with a person suffering from a disfiguring and potentially mortal disease, and at the same time get a sense of comfort out of the way she (the headdress and *tricolore* earring suggest we are looking at a woman) is tucked into her bed, and the manner in which the draughtsman took care with rendering the stripes of the scarf and a curl that escapes the head-cloth, allowing the eye to be distracted from the disease. Upon further contemplation, the oddities of the drawing come to the fore: the unveiled legs are small in relation to the head and neck; they are crossed in a distorted way and seem strangely detached from the upper trunk. It is a composite image, loosely held together by the sheet running diagonally through the watercolour hiding the suture between the body parts. The division is underlined by small inscriptions placed at right angles to each other and twice naming two distinct diseases: *Variola and Psoriasis*. What are we to make of this? Are we looking at a patient suffering from both smallpox and psoriasis, or is this rather a diagrammatic image – economically using the sheet of paper and demonstrating one disease on a head and another on legs?

According to an inscription on the verso, the watercolour was drawn by the Scottish anatomist Robert Carswell in 1829 at the Hôpital Saint-Louis in Paris. This 17th-century hospital had been transformed after the French Revolution into the first clinic dedicated specifically to the cure and study of skin diseases and was the major site for the formation of dermatology in France. Jean-Louis Alibert, who founded the dermatological school there, had from 1806 produced a set of publications on skin disease, including many that depict diseases in a portrait-like manner. What is particular about Carswell's

Variola and Psoriasis
Paris, 1829
Robert Carswell (1793–1857)
Watercolour on paper, 45 x 55.5 cm
UCL Library Special Collections. Fd 353 [21.]

medical watercolours, though, is that the anatomist himself made the drawings. The image in question here is only one of almost a thousand delineations (as the author himself called them) by Carswell kept in UCL Special Collections today and mostly depicting morbid phenomena observed upon autopsy. Carswell produced them in various Parisian hospitals, after his hiring by the University of London as England's first professor of Pathological Anatomy, in order to provide visual material for his teaching.

I am looking at this image as an art historian with a research interest in the history of skin and its representation, and it is one of my favourites among Carswell's watercolours because it doesn't cease to move and puzzle me. As it sits uncomfortably between a portrait of a patient and a portrait of a disease, it challenges assumptions about portraiture and prompts questions about the ways in which we engage with images of faces.

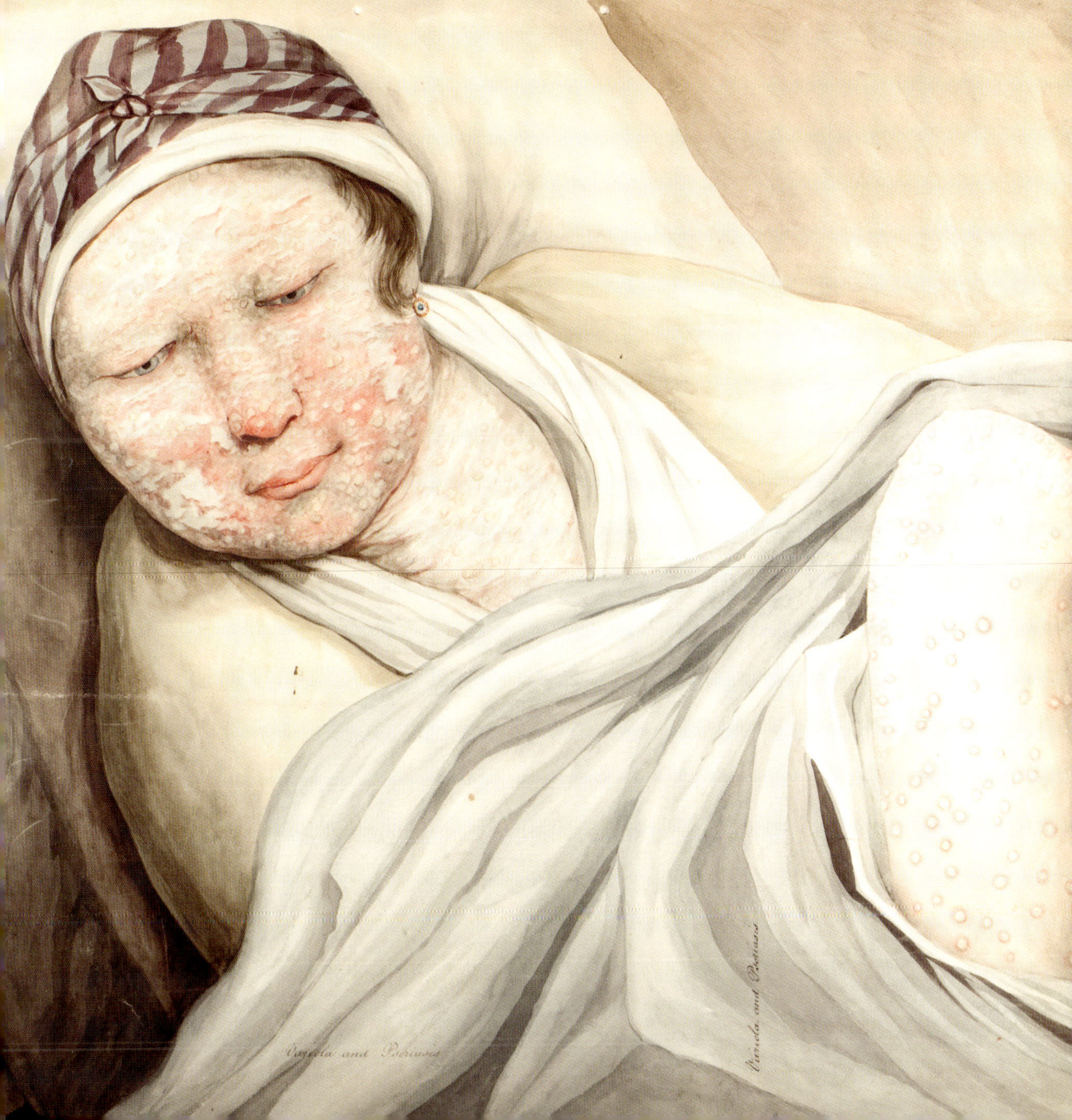

The Jar of Moles: Twitter celebrities and visitors' favourite

Professor Claire Warwick

Department of Information Studies

FOR ME THE CHOICE of an object from UCL's collections was remarkably easy to make. It had to be the moles: high-profile social media darlings, adored by the visiting public. It now seems almost impossible for me to give a lecture about digital humanities, social media or public engagement without the moles worming their way into it somehow.

Yet it was not always thus. For many years the moles – about thirty-two of them, we think – lived a life of obscurity, packed together in a jar, with only each other to talk to. Exactly what they discussed we shall never know.

Then came the move of the Grant Museum to its new site. As part of this project we in the UCL Centre for Digital Humanities collaborated with UCL Museums and the Centre for Advanced Spatial Analysis to create an entirely new form of digital interpretation for the museum – QRator (www.qrator.org). Using static iPads as interactive labels, visitors can respond to questions posed by the Grant's curators. We were delighted to find that visitors not only enjoyed doing this, but they also used the iPads to tell us about what they liked about the museum: and it became very clear, very quickly, that what they liked was the moles.

This was the opportunity the moles had been waiting for. Like many new celebrities they were keen to make their opinions heard and to connect with their adoring public on Twitter. Here, as @GlassJarofMoles, they have proved articulate and quirky, unafraid to court controversy, champion causes in which they believe, or tell jokes. In other words they have become a Twitter sensation. They were one of the stars of the #MuseumMascot day, a Twitter conversation involving creatures alive, extinct, and at various points between, from museums all over the world.

So far, so whimsical. Yet there is a serious side to the moles. Their prominence is evidence of the profound changes in the relationship between museums, their staff and visitors that digital technologies and social media can bring about. The Grant's curators now know far more about what visitors like, and want to see. As a result they can respond to visitor demand in a way that was previously impossible. QRator has given visitors a voice they previously lacked.

The use of Twitter by the moles, and other mascots, allows museums to make new connections with visitors and to maintain them after physical visits have ended. Children, in particular, love the idea of tweeting to an animal, and the medium allows the museum to inform visitors in a friendly non-didactic way. It allows visitors to reach through the virtual glass case and interact with the exhibits that inhabit them. The moles have become symbolic of the way that academics, technologists, museum professionals, and, yes, their exhibits, can connect with the public in a way that would have been impossible in a purely analogue world. It is no surprise, then, that they are my, and so many other people's, most memorable UCL museum object.

Glass jar of moles, Talpa europaea
Glass, moles, bakelite lid, height 30 cm
Grant Museum of Zoology and Comparative Anatomy UCL
LDUCZ-Z2754

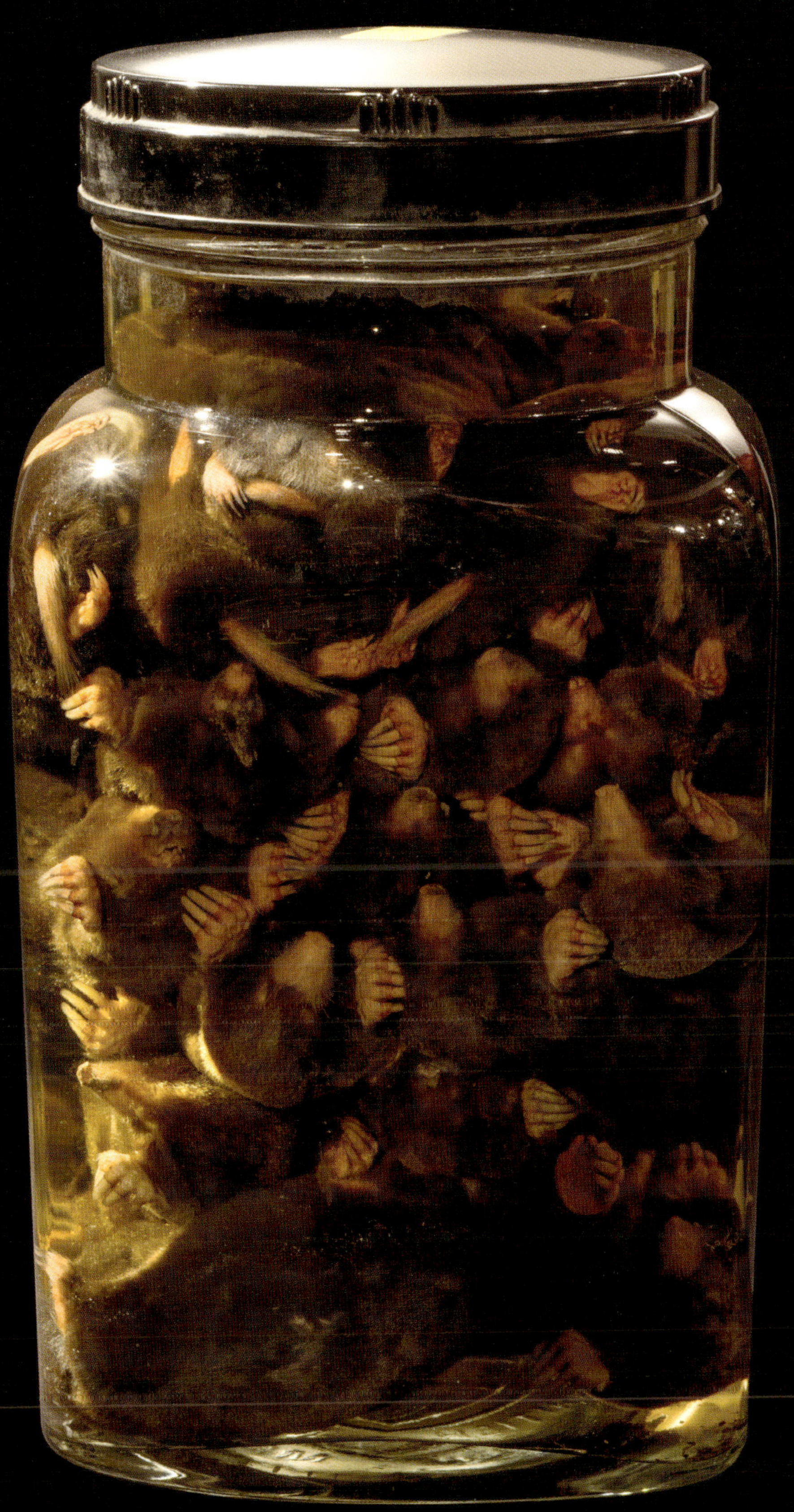

Welcome to Elizabethan London

Professor Helen Hackett

Department of English

O N 14 JANUARY 1559, a young woman of twenty-six processed through the streets of London to be crowned at Westminster Abbey. As Queen Elizabeth I she would become one of England's longest reigning and most renowned monarchs. Her journey to her coronation, cheered on by loyal subjects, was remarkably like a modern royal event, including crowd-pleasing moments:

> How many nosegays [small bouquets] did her grace receive at poor women's hands? How oft-times stayed she her chariot, when she saw any simple body offer to speak to her grace? A branch of rosemary given to her grace with a supplication by a poor woman about Fleet Bridge, was seen in her chariot till her grace came to Westminster.[1]

What was London like when Elizabeth processed through it? We can gain a vivid idea from the first surviving map of the city, engraved by Franz Hogenberg in 1572 using sources from the 1550s. Elizabeth set off from the

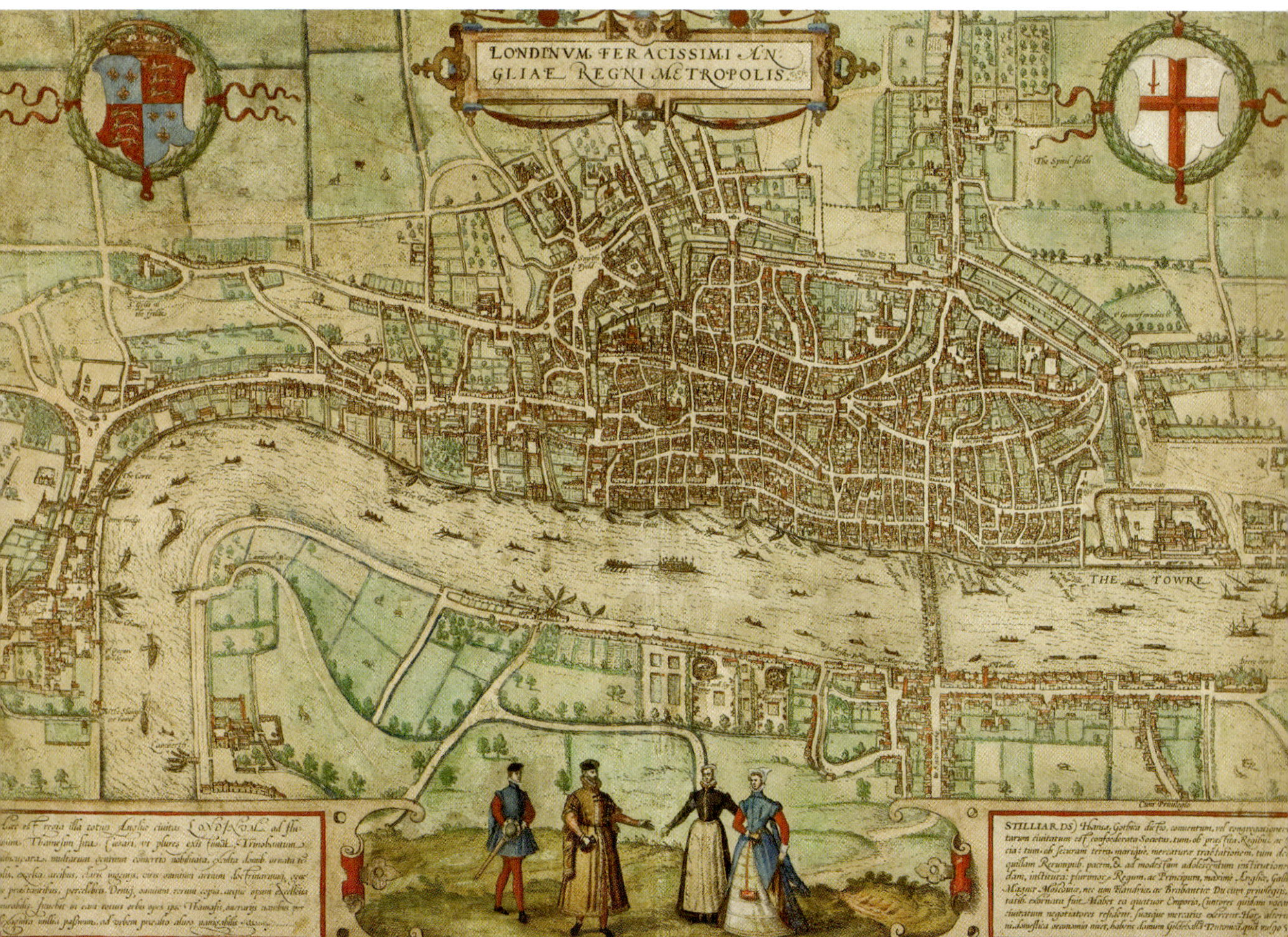

Tower of London in the east (to the right of the map). She went west through the densely built City of London, viewing pageants along the way; then followed the Strand, the road curving round the bend of the River Thames, to arrive at Westminster (at bottom left of the map) for her coronation.

Elizabethan London comprised two cities: the City of London to the east, the commercial centre; and Westminster to the west, containing the court and offices of government. The medieval walls of the City are visible on the map, surrounded by striking amounts of rural green space, all now absorbed into our present-day metropolis. Yet we can also see how London was already spreading beyond its walls. Over the course of Elizabeth's reign it would almost double in size to reach a population of 200,000.

On the south bank of the Thames we can see two circular structures, pits for bull- and bear-baiting. They were in this fringe location, outside the boundaries and jurisdiction of the City, because the City authorities strongly disapproved of such idle activities. By the end of Elizabeth's reign they were joined on Bankside by the equally disreputable playhouses, including Shakespeare's Globe. Only one bridge, London Bridge, linked the two riverbanks at this time; boats were the other means of crossing, and the Thames was also the main thoroughfare of the city, as we can see from the many vessels busily plying the water in Hogenberg's map.

It is surprising to learn that Hogenberg probably never visited London. This was one of hundreds of maps of world cities that he produced for a volume published in Germany. He got some things wrong: the Tower of London is fairly unrecognisable; and Old St Paul's (right in the centre) by this time had no spire. However, Hogenberg had good sources, and the map's level of detail encourages us to think that if we time-travelled to Elizabethan London we could find our way around. A group of amiable Londoners in the foreground welcomes us to their city.

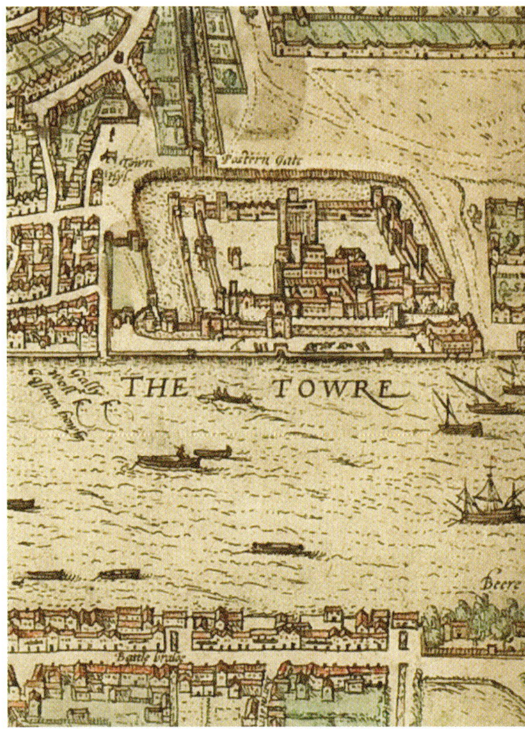

---

1. Germaine Warkentin (ed.). *The Queen's Majesty's Passage and Related Documents. Tudor and Stuart Texts* (Toronto: Centre for Reformation and Renaissance Studies, 2004), p.94.

Plan of London, from Civitates Orbis Terrarum (vol. I)
1572
Franz Hogenberg (1549–90); Georg Braun (born 1542)
Engraving with hand colour, 39.5 x 52.4 cm
UCL Art Museum
LDUCS_4787

Seeing is believing

Professor Sir John Tooke

Vice-Provost (Health)

VISION IS A HUMAN being's most precious sense and also the most sensitive. Since the beginning of the science and art of medicine, doctors have relied upon their powers of observation to diagnose countless illnesses. From characteristic rashes and the colour of the skin to facial expression and body movement, vision provides the doctor with evidence of diseases as varied as chicken pox, jaundice and Parkinson's disease.

Yet our vision limits us to a view of surface features. The development of x-rays enabled doctors for the first time to view what lay beneath the skin, where most pathology occurs. From that point the advance of medical science has accelerated, with modern imaging a crucial, largely non-invasive addition to the doctor's diagnostic (and therapeutic) skills.

For these reasons I view the x-rays taken by Norman Collie, professor of Organic Chemistry, who came to UCL in 1887, as a genuine landmark in the history of medicine, ushering in an era of remarkable medical discovery. Collie had completed his PhD in Würzberg and happened to be there in December 1895 when

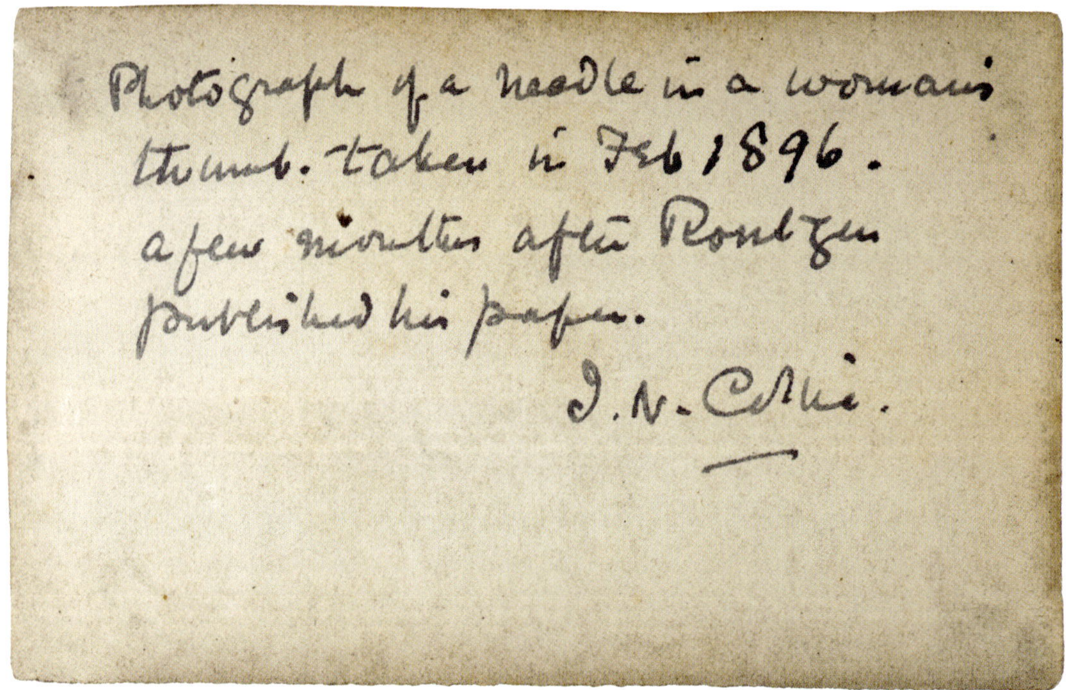

X-ray
1896
15.7 x 13 cm
UCL Chemistry Collections
LDUSC-CHEM-105

William Röntgen, German physicist, accidentally discovered the penetrating power of x-rays. When Collie returned to UCL he carried out a number of tests, often x-raying animals. But it was the x-ray he performed in February 1896, on an actress who had a needle stuck in her thumb, that marks the turning point in the use of this new technology. This radiograph was the first use of x-rays for clinical purposes in the United Kingdom, if not in the world.

Since those early days imaging science has grown enormously at UCL, with several groups that are international centres of excellence in their own right. The Centre for Advanced Biomedical Imaging develops novel imaging technologies using a wide range of physical principles including photo acoustic and magnetic resonance. The Wellcome Trust Centre for Neuroimaging allows the interrogation of the function of the brain as well as structure. These advances rely on the close co-operation of scientists drawn from the physical sciences, engineering and computing as well as medicine, an interdisciplinary approach which is a hallmark of UCL's approach to the resolution of really challenging problems, heralded by Collie's departmental origins.

Medicine has moved a long way since the pioneering days of Norman Collie. Although when reflecting on medical advance we tend to think of new drugs and surgical methods, none of these would be possible without the huge developments in medical imaging.

Mountain climbers say that the view from the summit of the world's highest mountains gives a different perspective on the world and indeed life itself. Collie, as befits a pioneer, was also a very accomplished climber and is credited with seventy-nine first ascents, from the Himalayas to the Alps, and named more than thirty peaks in Asia, Europe and America. It is perhaps then little surprise that he pioneered the use of x-rays in a clinical setting and in so doing opened medical eyes to the mysteries that lie beneath the body's surface.

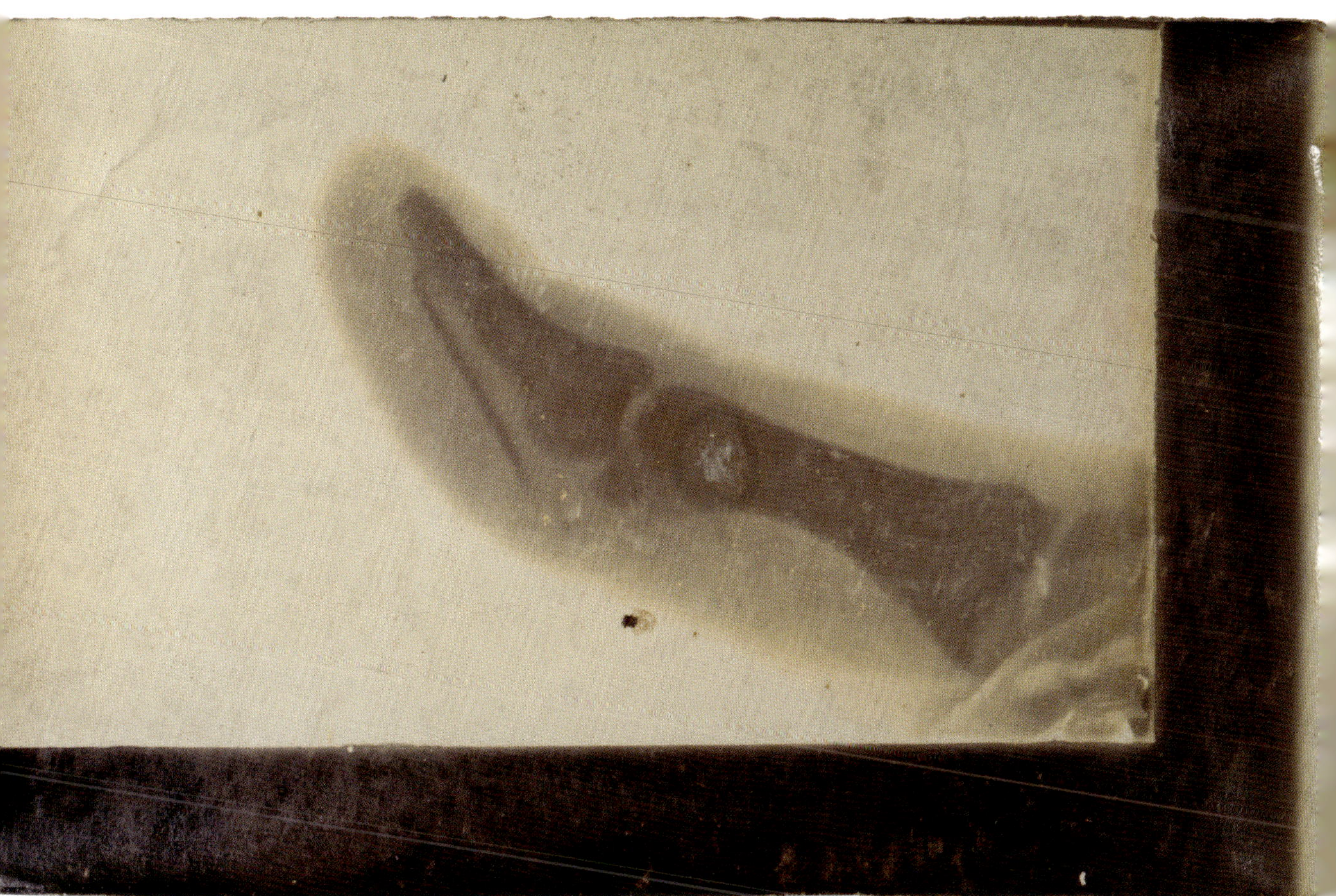

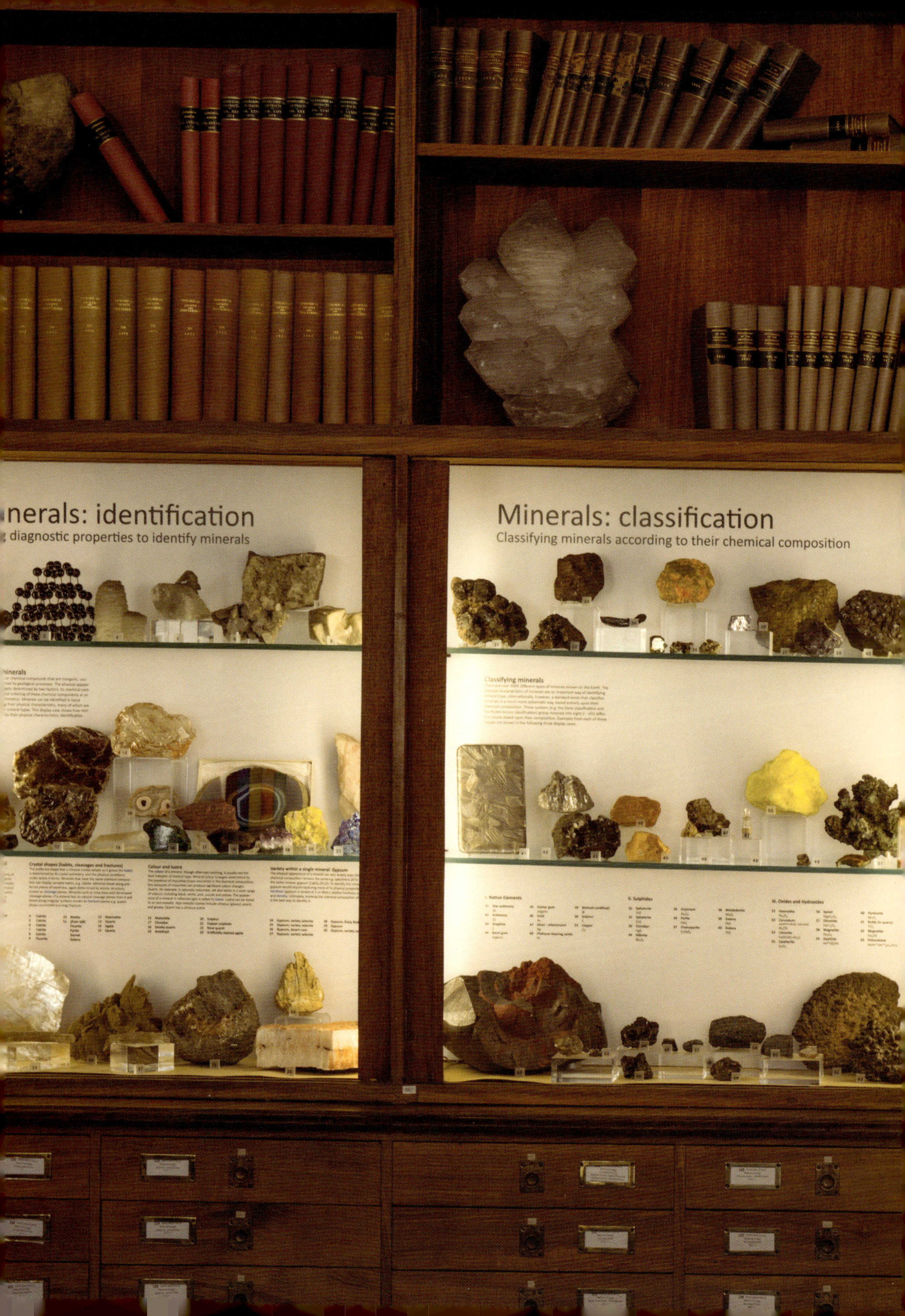

nerals: identification
diagnostic properties to identify minerals
Minerals: classification
Classifying minerals according to their chemical composition

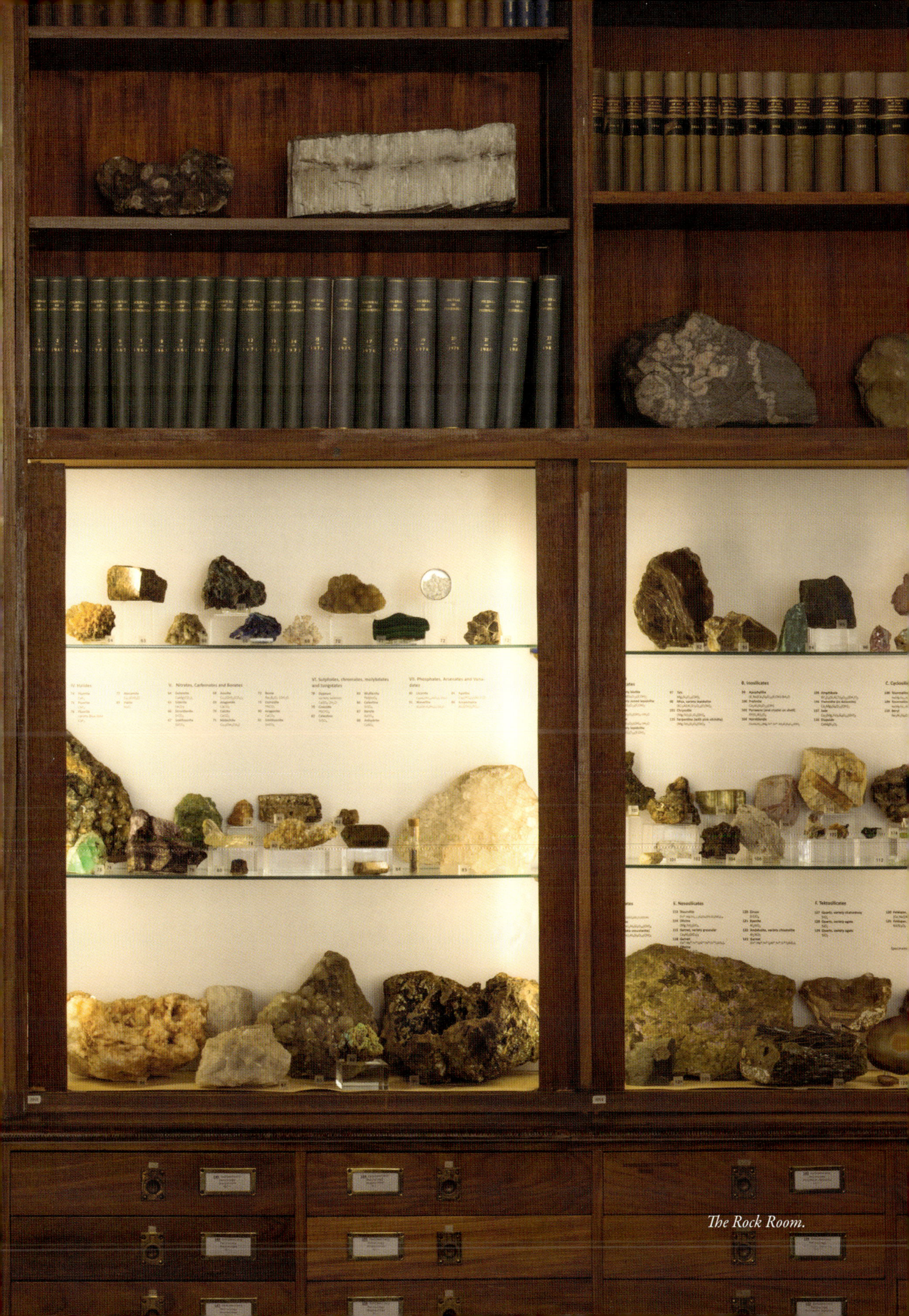

The Rock Room.

A casserole of Egyptian blue

Professor Thilo Rehren

UCL Qatar

Serves plenty – the highlight in any dining room or banquet

Ingredients

1 bucket of clean white sand
½ bucket of good limestone (crushed marble or sea shells will also do)
¼ bucket of flower of copper (or burnt bronze scrap from the local foundry)
1 ladle of natron
large casseroles or saggars that can be stacked on top of each other
large wood-fired kiln

Preparation

Grind the sand, limestone and flower of copper very finely (until you can no longer feel any grains between your fingers). Add the natron, and mix well until evenly grey. The mixture will keep for years when stored in a dry place. For use, dampen a large quantity with some water, beer or gum arabic and knead thoroughly. Form into bean-sized pearls and put them on a tray in a sunny place to dry.

Baking

Take several large ceramic casseroles or saggars; coat liberally on the inside with a lime wash to prevent contents from sticking. When dry, fill the vessels to the top with the prepared pearls; be careful not to squash them. Stack saggars in a kiln one atop the other, covering the uppermost one with a lid. Fire to a dark red glow for a couple of days, or until the content is bright blue (beware the heat when removing a sample). When ready, remove saggars from kiln, allow to cool, and retrieve the Egyptian blue pearls as needed.

Serving suggestion

Used best as finely ground pigment with ochre, cinnabar, lead white and carbon black for fine wall paintings on a light plaster coat. Mix with extra lime for lighter shades of blue; use coarse powder for dark blue. Point out to your guests that this is the only blue pigment worth having, after lapis lazuli.

Special tip

Dampen the finely ground powder with gum arabic or egg white and shape into beads, pendants or small vessels and fire a second time; makes fantastic trinkets and special gifts. Inlay with yellow for colour effect.

Base fragment of an industrial pottery vessel
Memphis, Egypt, Ptolemaic period, 323–30 BC
Clay, 13.5 x 3 cm
UCL Petrie Museum of Egyptian Archaeology
LDUCE-UC47311

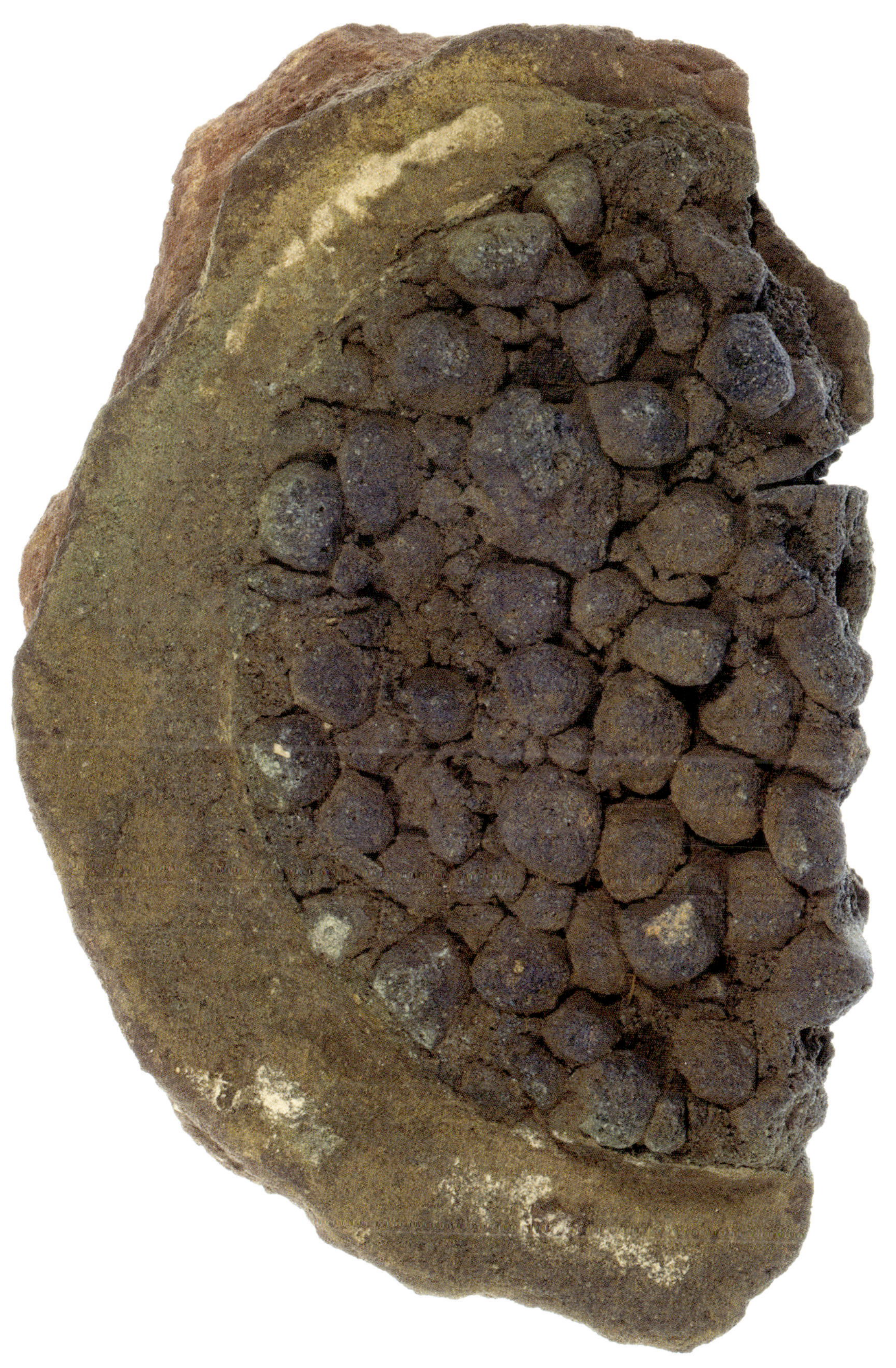

The unmaking of a king, the making of Shakespeare

Dr Chris Laoutaris

Department of English

THIS PRINT CAPTURES the unmaking of a king. One of a series of Shakespeare-related rare artworks held in the UCL Art Museum, it was commissioned in 1801 by the printseller John Boydell (1720–1804). Engraved by Benjamin Smith, it records the scene in Shakespeare's *Richard II* during which the king of the play's title relinquishes his throne to his usurping cousin Henry Bolingbroke. In a poignant reversal of the sacred rites of the coronation ceremony, Richard divests himself of the most potent emblems of his regal authority: the crown and sceptre. The print's symmetrical composition, bisected by the contested crown, mirrors Shakespeare's conception of the breath-taking moment when the balance of power is momentarily suspended before its tragic tipping point:

> Here, cousin, seize the crown. Here, cousin,
> On this side my hand, and on that side thine.
> Now is this golden crown like a deep well
> That owes two buckets, filling one another,
> The emptier ever dancing in the air,
> The other down, unseen and full of water.
> That bucket down and full of tears am I,
> Drinking my griefs whilst you mount up on high.[1]

There is another, hidden, story to be read in this engraving: that of Shakespeare's developing status as national poet and dramatist in the late 18th century. Smith's print was part of a grand project orchestrated by John Boydell for the founding of the world's first Shakespeare Gallery. This idea took shape in November 1786 at a dinner party hosted by John Boydell's nephew, Josiah Boydell (1752–1817), at which 'it was remarked that the French had presented the works of their distinguished authors to the world in a much more respectable manner than the English had done'.[2] Spurred on by these developments in continental literary culture and art, John Boydell resolved to energise British history painting, and exploit a potential international market for English prints, by commissioning new illustrated editions of Shakespeare's plays.

On 4 May 1789 John Boydell and his nephew opened the Shakespeare Gallery in London's Pall Mall, showcasing the talents of the artists involved in this ambitious publishing venture. The gallery drew attention to the work not only of British artists, like Joshua Reynolds and Benjamin West, but of painters and engravers who had been attracted to England's art scene, such as the Swiss artist Henry Fuseli, the Swiss-Austrian portraitist Angelica Kauffmann, and the Florentine engraver Francesco Bartolozzi. The Boydells' mission to

Shakespeare. King Richard the Second. Act IV, Scene I after Brown, M.
Date unknown
Smith, Benjamin (1775–1833)
Engraving, 49.8 x 63.2 cm
UCL Art Museum
LDUCS_4619

raise the reputation of British art was therefore not an insular one, but was realised through cross-national collaboration and exchange.

When I take students to the UCL Art Museum to see the engravings which grew out of John Boydell's vision, I am bringing them face-to-face with the surviving remnants of a project which contributed, in significant ways, towards the making of the globally recognisable face of the Shakespeare we all know today.

<hr>

1. William Shakespeare (ed. Charles R. Forker), *Richard II* (Arden Shakespeare, 3rd revised edition, 2002), act 4, scene 1, lines 182–89.
2. Joseph Farington (ed. Kenneth Garlick, Angus Macintyre et al.), *The Diary of Joseph Farington volume III* (Yale University Press, 1979).

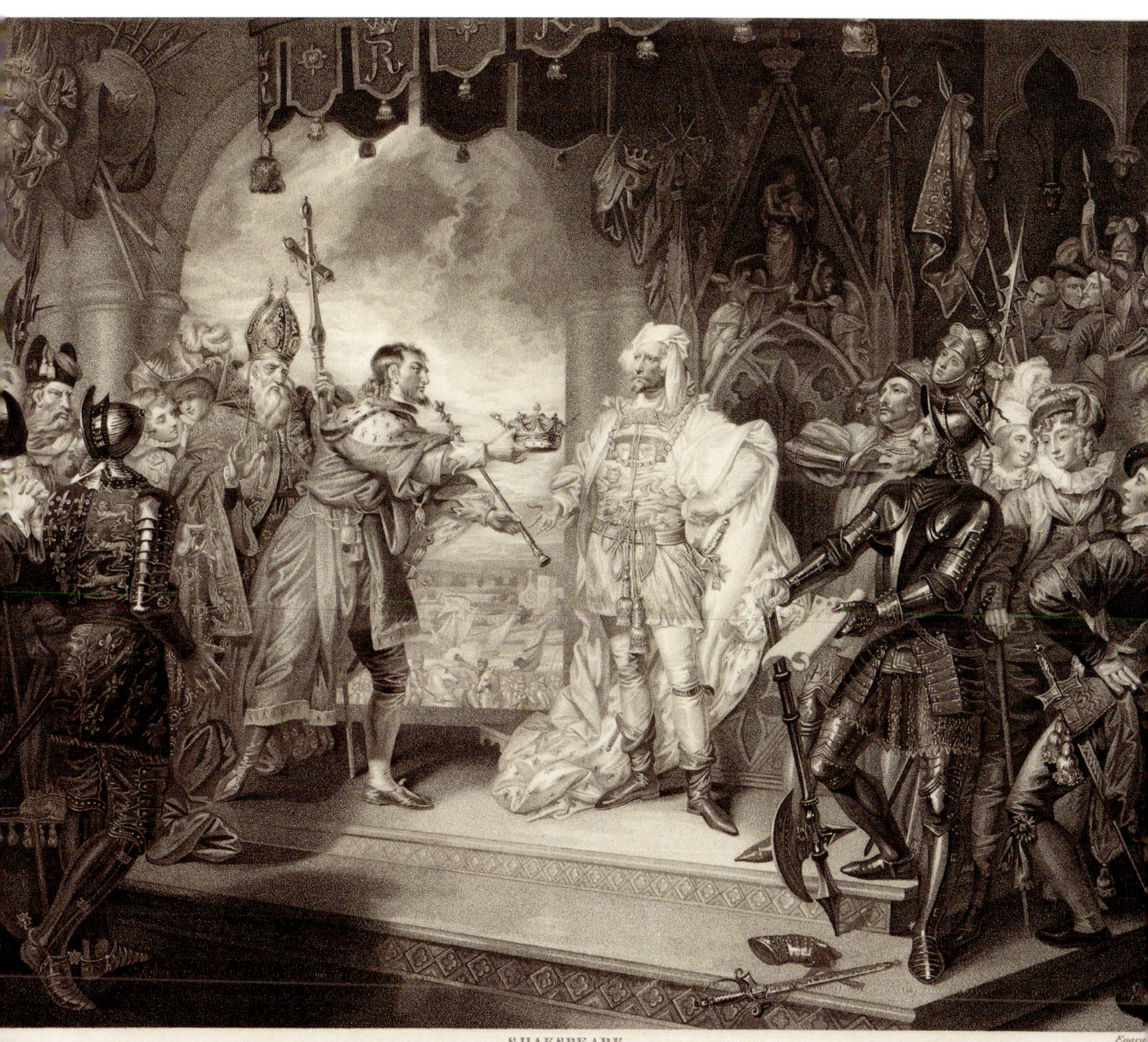

Scholarship, extinction and the Irish Elk

Professor Mary Collins

Dean of the Faculty of Life Sciences

THE IRISH ELK SKULL, with its magnificent antlers, used to guard the stairs of the Zoology department at UCL. You walked past it to get to the Grant Museum, which in those days was on the top floor of what is now the Medawar Building. In 1984 I was a junior researcher trying to bring the new gene-cloning techniques to the department. The professor would only buy equipment from UK companies; one machine always jammed shut and had to be opened with a spanner. The reagents we used were homemade. Distilling highly toxic phenol, a simple process in well-equipped chemistry labs, was always challenging for us biologists. Seminars were a lively and slightly frightening forum to present work in progress. At 3.30 p.m. every day there would be tea and biscuits and the professor would hold court.

Part of the department was 'the professor's hangers-on', including the lab where I worked. At the other end of the corridor were experts on amoebae, ecology, frog skin and an elderly and renowned Darwin scholar. There was also a man who walked round with a pet ferret on his shoulder (I never found out what he did). In my flurry of competition with groups in Boston and Los Angeles I did not pay them much attention. Towards the end of my time in Zoology I became frustrated with what I thought of as eccentric working

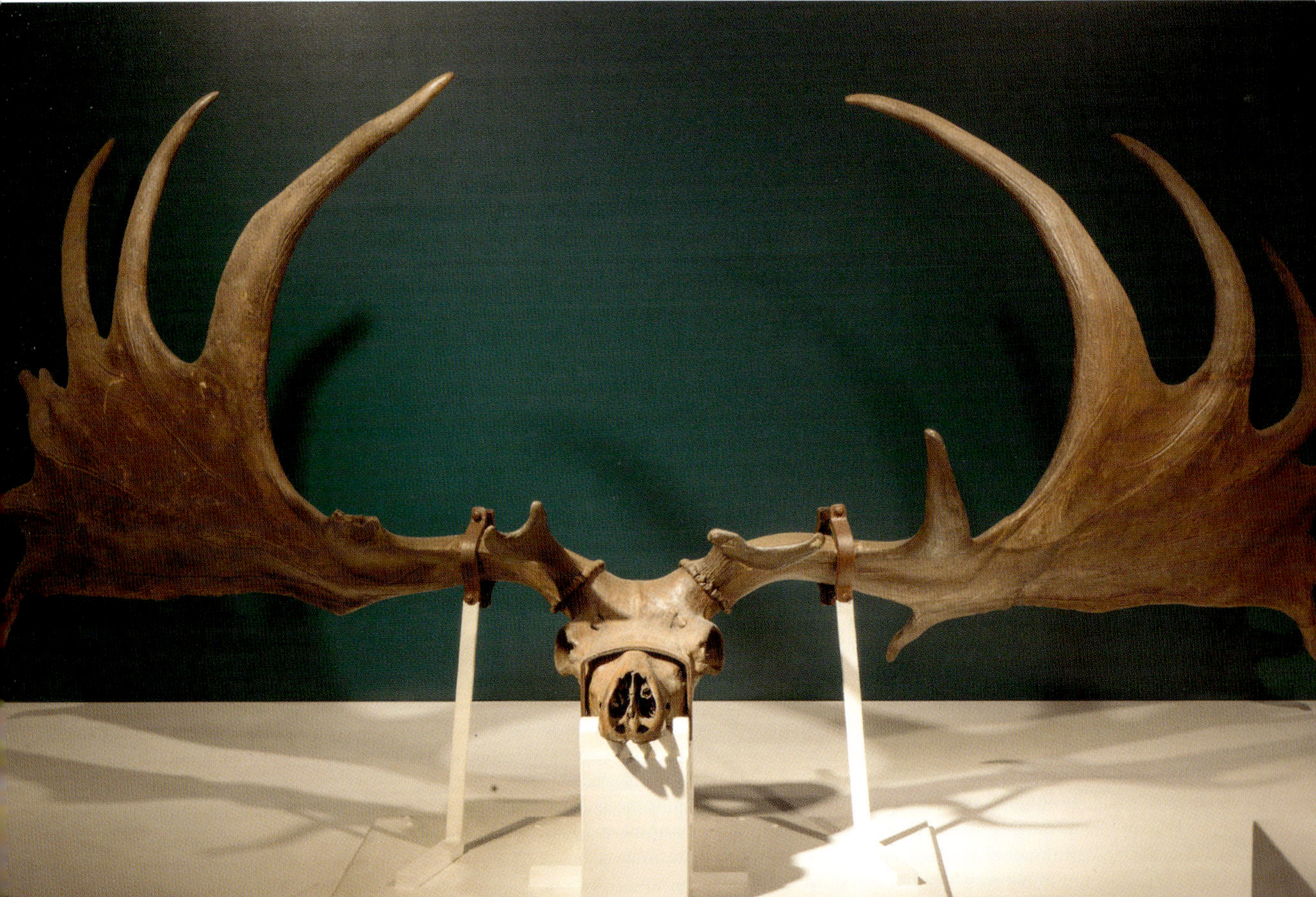

conditions. For my next job I decided to go to a brand new Institute at MIT in Boston. The Grant Museum was a good place to concentrate and that was where I wrote my fellowship applications.

Working in the United States was very instructive. I learnt that there were no such phrases as 'too expensive' or 'too much work'. Seminars were guarded affairs. Everyone worked on aspects of molecular cell biology. The maintenance team did a cabaret act at Christmas, including 'The Director Rap' – 'We know he's smart, he won the Nobel Prize'. The director was the most competitive person I've ever met, even in softball games. The building was state of the art, designed to foster interactions; there was a Frank Stella painting in our corridor. Junior researchers at the Institute worked all night.

My zoologist friends tell me that the Irish Elk isn't an elk, or a moose as my Bostonian colleagues would have had it, and it isn't particularly Irish. It was a close relative of the fallow deer, and was possibly the largest deer that ever lived. It roamed across the whole of northern Eurasia during the Pleistocene and early Holocene, disappearing from the fossil record around 8,000 years ago. Irish Elk appear in the famous hunting scene paintings in the caves at Lascaux, France; and although there are still those who take a contrary view, it seems inescapable that humans – those questing, innovative, competitive creatures – played a key role in its demise.

Subfossil skull and antlers of giant deer, Megaloceros giganteus
117 x 261 x 152 cm
Grant Museum of Zoology and Comparative Anatomy UCL
LDUCZ-Z800

Silica Aerogel

Professor Mark Miodownik

Department of Mechanical Engineering

AEROGELS DO NOT LOOK like anything else you have ever seen. If someone told you that they had been discovered in a crashed spaceship, you would believe them; everything about the material is alien. Aerogels have the ability like nothing else to compel you to search your brain for some excuse to get involved with it. Like an enigmatic party guest, you just want to be near it, even if you can't think of anything to say.

Aerogels were first discovered by Samuel Kistler in 1931. He had been interested in understanding the structure of gels and in proving whether a gel contained a continuous solid network of material. He chose to do this by finding a method to remove the liquid from a gel without collapsing the solid pores, and in succeeding created the most highly porous material in the world. First he did it with silica, a pure form of sand, and this is what we see here. It is a glass foam whose nano-structure contains up to 99.8 per cent air, making it one of the world's lightest solids. This sample was made by Steve M. Jones, at NASA Jet Propulsion Laboratory, as part of the NASA Stardust Mission, which involved sending a spacecraft containing a large piece of aerogel on a close approach to the comet Wild 2 to collect space dust and return to Earth. Aerogel was ideal for the mission because its ultra-fine foam could gradually decelerate and capture dust particles in pristine condition. The return capsule containing the stardust successfully re-entered the Earth's atmosphere and landed on 15 January 2006.

Silica aerogel appears to be much more invisible than glass despite being less transparent. This is because there is no hint of reflection on its surfaces, giving it the appearance of not being fully solid. Its azure colour is not due to any pigmentation, but is caused by the same phenomenon that gives colour to our Earth's atmosphere, namely Raleigh scattering of light. So it really is the ultimate blue sky material.

Silica aerogel sample
NASA Jet Propulsion Laboratory, 2006
Steve M. Jones
Silica nano foam, 6.5 cm
UCL Institute of Making Materials Library

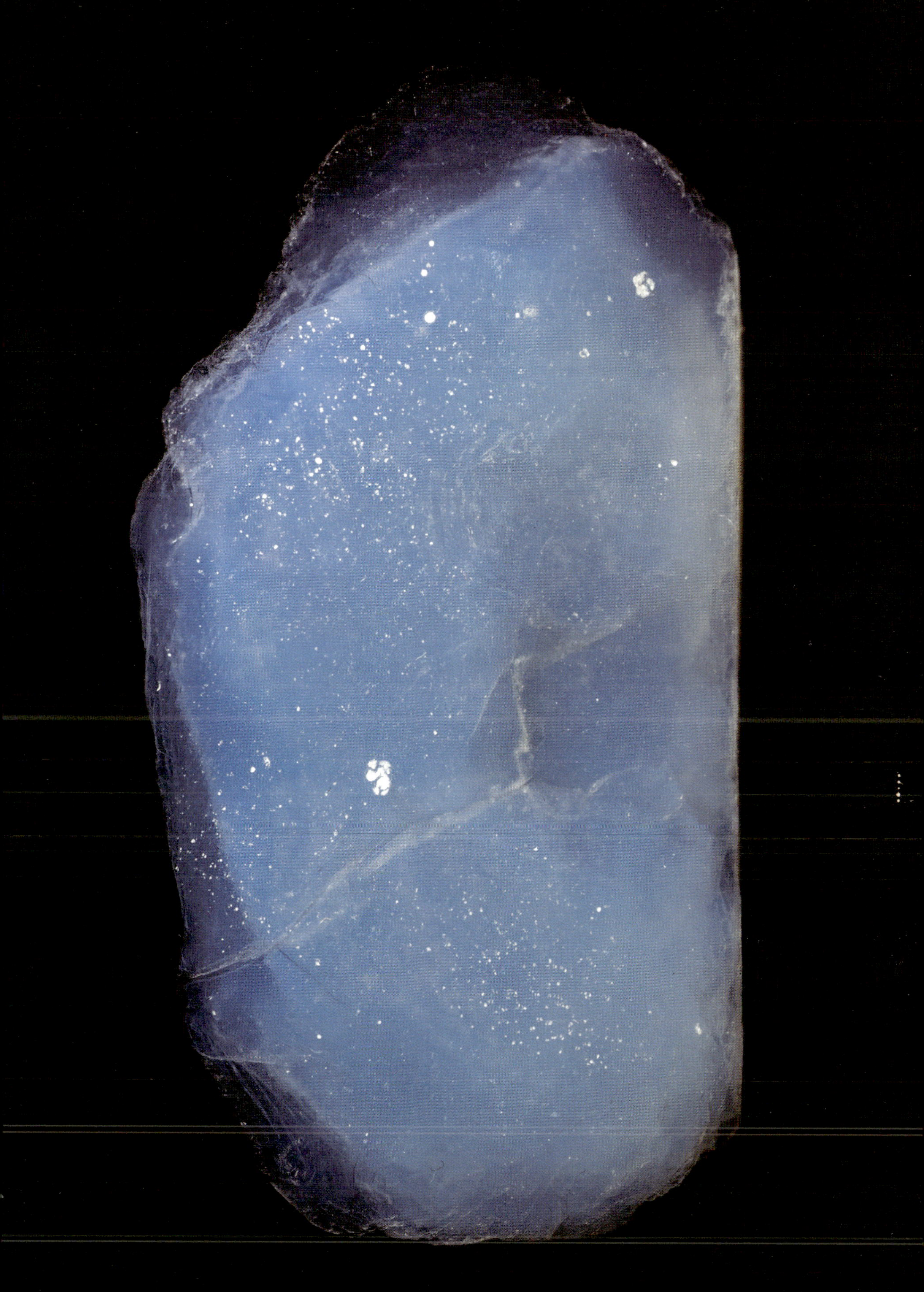

Honey for the soul and for scholarship

Professor Gesine Manuwald

Department of Greek & Latin

UCL SPECIAL COLLECTIONS owns a 16th-century edition of the work of the Roman philosopher Lucretius. Why is it so spectacular?

A scholar of Latin is familiar with modern editions of Lucretius' poem 'De rerum natura' (literally 'On the nature of things'). In the first century BC, Lucretius followed the doctrines of the Greek philosopher Epicurus, who argued that the natural world consisted of atoms and had not been created by gods. He thought that gods were not interested in human beings; hence there was no need for humans to fear them. Such issues are set out by Lucretius in the six books of his poem, whose beautiful lines he compared with honey put around bitter medicine, so that readers could absorb the difficult philosophical doctrines more easily.

In the Middle Ages, Lucretius' work was virtually forgotten until the humanist Francesco Poggio Bracciolini found perhaps the last surviving handwritten manuscript in a German monastery. The very first printed edition appeared in 1473.

Hence the early-16th-century edition owned by UCL is one of the earliest; it is a second edition from 1514, triggered by the impact of the first edition of 1511. It not only contains the text of Lucretius' poem, but also an extensive printed commentary by the Italian humanist Giovan Battista Pio (1460–1540). This copy also includes handwritten notes. The extent to which Renaissance scholarship engaged with Lucretius' poem can be gauged from the fact that there are only a few lines of Lucretius' text on each page, while the entire area around them is filled with commentary, a writing style adopted from medieval manuscripts.

The commentator's position is obvious from the title page (pictured): the long title highlights the merits of the commentary, promising to explain all difficulties and to provide parallels from Greek and Latin authors, on a larger scale than any previous works. Pio is quite right in saying so, since his was the first proper commentary on the whole of Lucretius' poem. The commentary discusses grammatical and linguistic points as well as historical, mythical and philosophical details, including various digressions; it is fairly objective and sets the text in the wider context of ancient and medieval philosophy.

So it is a great feeling for Latinists when they can get a copy of one of the first extended scholarly efforts on an ancient text from their university library. They can relate to the beginnings of scholarly treatment of those texts, since this comes close to what academics do today. It is also reassuring to see that at least the efforts of this scholar were read: this second edition of the Italian humanist's work was printed in France by the famous printer Jodocus Badius Ascensius (whose printing office can be seen in an emblem in the middle of the title page), and indications of former owners at the front indicate that the book was in use in France before it was acquired by UCL in 1938.

In Carum Lucretium poetam commentarii a Ioanne Baptista Pio editi
Paris, 1514
Titus Carus Lucretius
Paper, bound in leather, 33 cm
UCL Library Special Collections
SRC Quarto 1514 L8

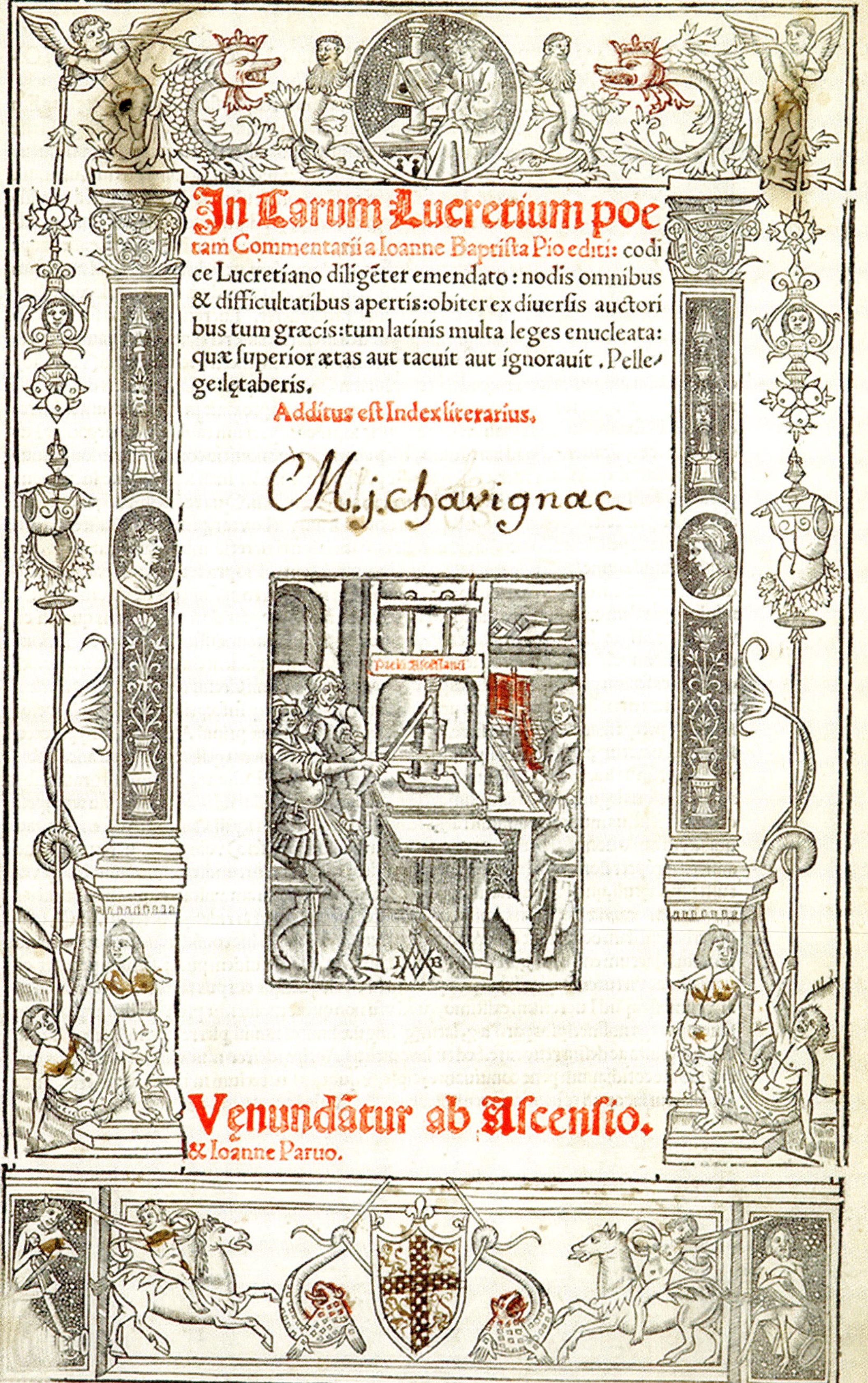

In Carum Lucretium poe

tam Commentarii a Ioanne Baptista Pio editi: codi
ce Lucretiano diligeter emendato: nodis omnibus
& difficultatibus apertis: obiter ex diuersis auctori
bus tum græcis: tum latinis multa leges enucleata:
quæ superior ætas aut tacuit aut ignorauit. Pelle=
ge: lçtaberis.

Additus est Index literarius.

M.j.chavignac

Venundatur ab Ascensio.

& Ioanne Paruo.

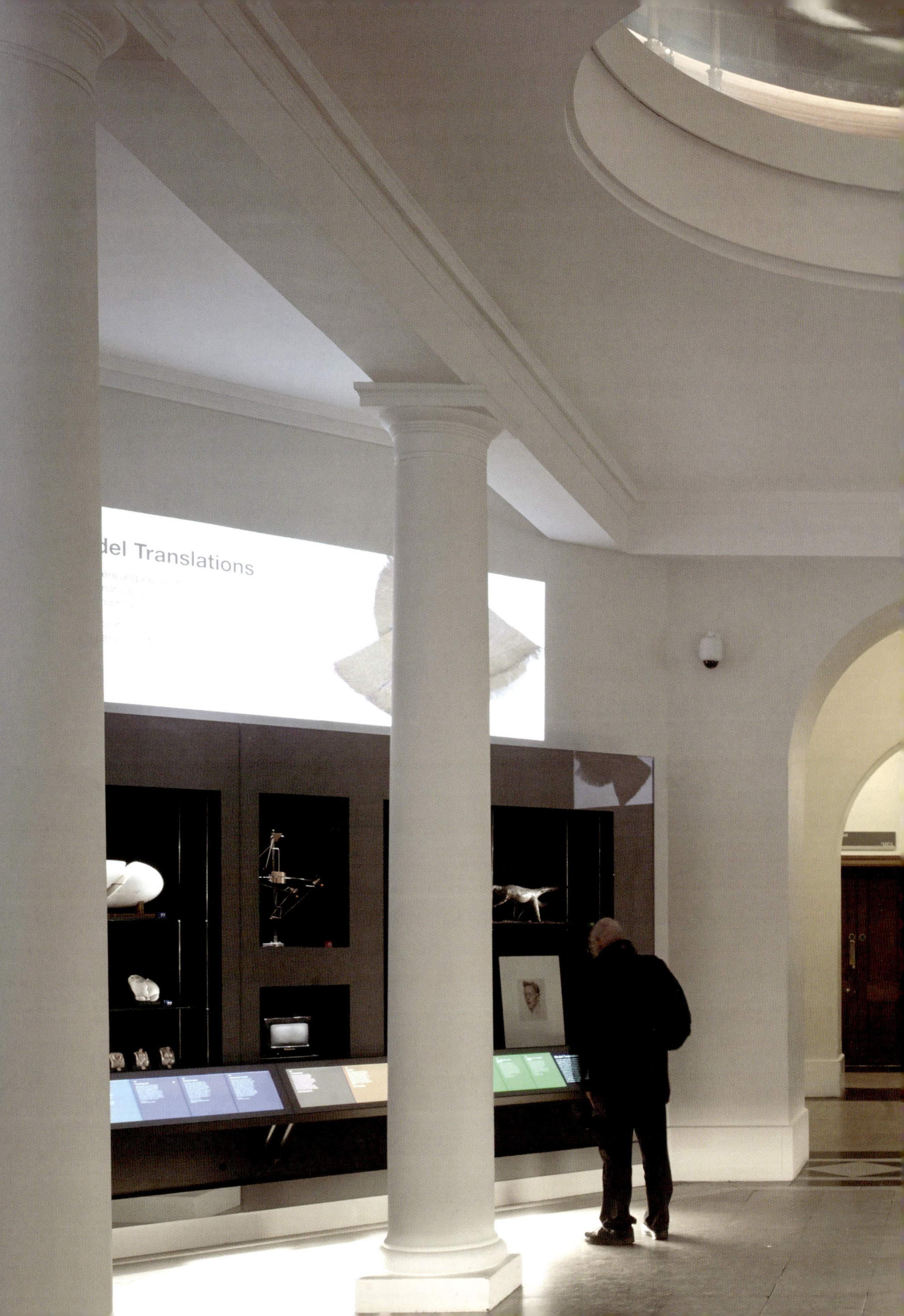
del Translations

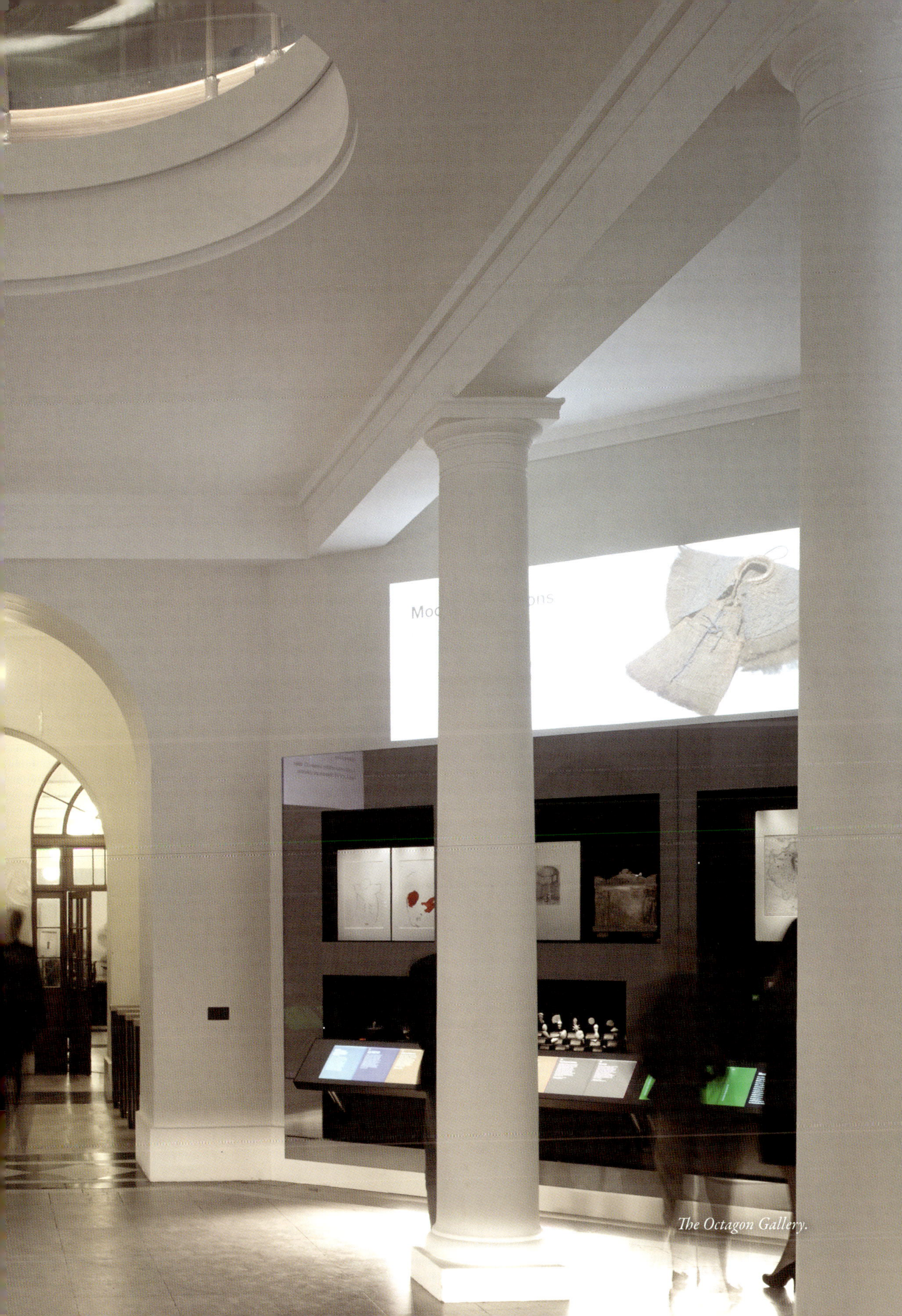

The Octagon Gallery.

A lute player on the Nile

Janet Picton

Institute of Archaeology

She stands ready, poised, lips softly rehearse praises to the Golden One.
Loveliest of all Hathor's famous beauties, naked she greets the rising sun.
Protector of Horus in the marshes (Lady of Turquoise, sister of noble Isis).
Lilies rise in dawn's mist dripping jewels. Rustling about you, green papyrus.

Isis. Lady! (Osiris's queen of night, mistress of heka, heroic god's mother).
I beg! Raise desire, sate, grant succour; stir the gleaming limbs of my lover.

Oh, she is flawless, lovely beyond compare, above all Egypt's radiant beauties.
Strengthen me Ptah, guide my hand, let me do justice to my craftsman's duty.

My love, Nefret, serves her ladies; fragrant lilies perfuming festive braids.
Slender golden curves, sleek skin dewed with effort as agile fingers played.

Naked as the dawn, her armour spell-bound beads and belt of cowrie shells.
Head bows on nodding neck, all awareness centred on long-learned spells

Have I shown my love to you? See! Her graceful throat trembles, breasts rise
to draw breath – praise song soars forth, dancing in the shimmering air and skies

Hidden in the reeds Taweret guards her fragile skiff, Geb's emblem at the prow.
Denizens of the marsh defend her until duty done, and silent lute laid down.

Hail, my lady of the marshes!

As teachers we say that we can not know how people of the ancient past thought, or what they felt. Sometimes we get a fleeting glimpse of understanding from the ancient texts that we are so fortunate to have from Egypt. One example is this poem from Papyrus Chester Beatty I.

We attempt to interpret this 'unguent spoon' with its exquisite carving of a young girl in the marshes at the edge of the Nile. The girl stands in a papyrus skiff playing a lute, naked except for a beaded belt and collar, wearing an elaborate hairstyle with a band of lilies at her brow. An enchanting scene, but what does it mean?

We make connections to the worship of the goddess Hathor and the young musicians known as the *nfrt*, 'Hathor's famous beauties'. The spoon is shaped like an *ankh*, meaning 'life': was it used in a ritual? We discuss the presence of two symbols of ancient Egypt – the papyrus and lily. We think about the types of wood available, and examine the distinctive craftsmanship of the man called the 'papyrus-fan carver' by scholars. We are no closer to knowing what the object meant to the ancient society that made and used it.

However, when we bring the two together, the lovesick narrator of our poem and the graceful young woman of the spoon, we can believe – just for a moment – that we have made a connection to the immediacy of life and love in Egypt 3,300 years ago.

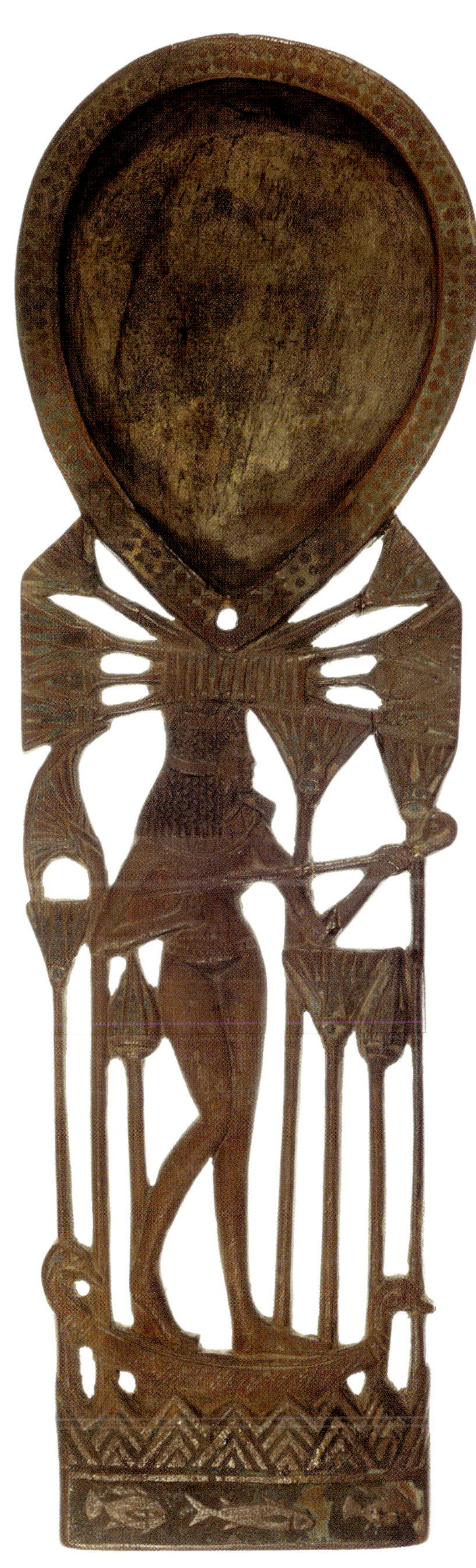

Cosmetic spoon
Sedment, Egypt, late Dynasty 18, about 1350 BC
Wood, 21 x 6.35 cm
UCL Petrie Museum of Egyptian Archaeology
LDUCE-UC14365

Life in glass

Dr Petra Lange-Berndt

Department of History of Art

DRESDEN-BASED GLASSWORKERS Leopold Blaschka and his son Rudolf are famous for their meticulous recreations of invertebrates dating from the late 19th century. I have always been drawn to these strange little creatures that are to this day unmatched in their craftsmanship. Animals like the tooth-finned or comb-finned squid have soft bodies that are difficult to preserve by drying or immersion in aseptic liquids. Furthermore, these organisms are beautiful to the eye but challenging to touch and smell: substances that are sticky and between states evoke repulsion and nausea in their human audience. This is why the Blaschkas experimented with traditional Bohemian glass-blowing techniques in order to convert slimy tissue into a durable and odorless substance. In doing so, they transformed chaotic organic life into hard scientific facts. Their objects resonate with both the art and zoology of that time.

To appreciate the tiny squid (measuring 13.5 centimetres) held by the Grant Museum of Zoology – one of thirty-three individual glass models – requires close observation as this object is too delicate to handle. Floating in water and connected to thousands of other organisms when alive, the model of this creature has been singled out and placed into a box in front of a monochrome black background. It has become an educational aid. The opalescent and slightly translucent glasswork evokes the slippery body that seems strangely animated by its detailed eyes and tentacles – the Blaschkas wanted to be true-to-life and kept animals in aquariums for observation in their studio. But any organic appearance of the squid is kept in check by the symmetrical design of the body. Leopold and Rudolf Blaschka worked at a time when evolutionary theories were widely debated; in this context the deep sea had received increased attention. Jena-based Ernst Haeckel for instance carried out field research describing about 4,000 new species of invertebrates. Publications like *Radiolaria* (1887) and the famous *Art Forms of Nature* (1899–1904) became influential for Art Nouveau design since the biologist and artist had transformed life forms into symmetrical ornaments for the accompanying illustrations. But this is not an innocent design: a specific standard of beauty defines one's place on Haeckel's imagined evolutionary tree – in this case a German oak – where the white male is dominating the rest of the creation. The naturalist, like the Blaschkas, is operating in the realm of popularized science.

But the meaning produced by the Blaschkas is not fixed for eternity. PhD candidate Pandora Syperek has managed to find a way to further destabilize their order of things. She has been carrying out research for her thesis on how objects on display at the Natural History Museum in London have been gendered. Syperek interestingly points out that some of the life forms in question present their audience with non-normative models of sexuality: there are hermaphrodites and colony-based animals and creatures that can spontaneously lose their limbs. In that sense the tiny crystallized glass bodies are Trojan horses that can unexpectedly oscillate from hard fact to subversion.

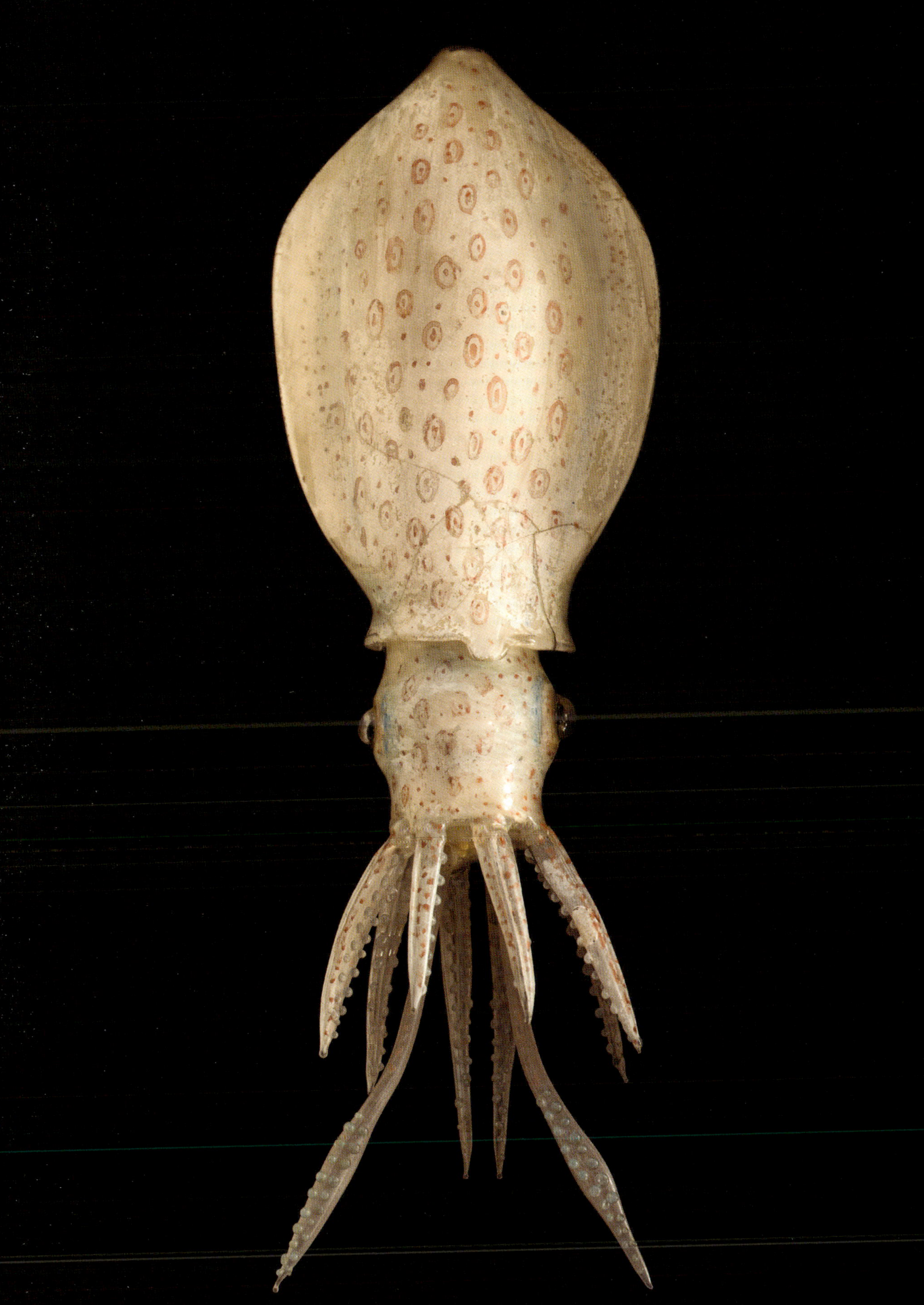

Digging a little deeper: The Greenough Society Members' Book

Professor David Price

Vice-Provost (Research)

MY BACKGROUND IN Earth Sciences has involved me in a lifetime of exploring what lies beneath the surface. In addition to finding endless inspiration in the study of our own planet and being intrigued by the many natural wonders in the UCL Geology Collection, I have always been fascinated by the members' book of the Greenough Society, which is the students' social club in what is now UCL's Department of Earth Sciences (formerly the Department of Geology). The society was founded as a dining club for members of the department, and was one of the very first student societies at UCL. It was created to celebrate and to enable 'Fieldwork, Friendship and Fun' and is still thriving today.

Reading the names of the society's early members, I am struck by how many of them were female, even in the days when UCL was not as well balanced as it is today. Notably, one of the founding members of the Greenough Society was Marie Stopes. She studied geology and biology as a student, and obtained a first-class degree after only two years at UCL. She then went on to obtain a PhD from the University of Munich after only a further two years of study. Marie and the society's other female members were pioneers, challenging the image of the quiet, domestic Edwardian woman, and forging new paths in the predominantly male world of higher education. But Marie achieved this perhaps more so than the others. She returned to UCL in 1911 as a lecturer in palaeobotany, and is still known for her work on coal and 'coal balls'. She was still a lecturer at UCL when in 1918 she wrote her famous book *Married Love*, following which she moved on to greater things!

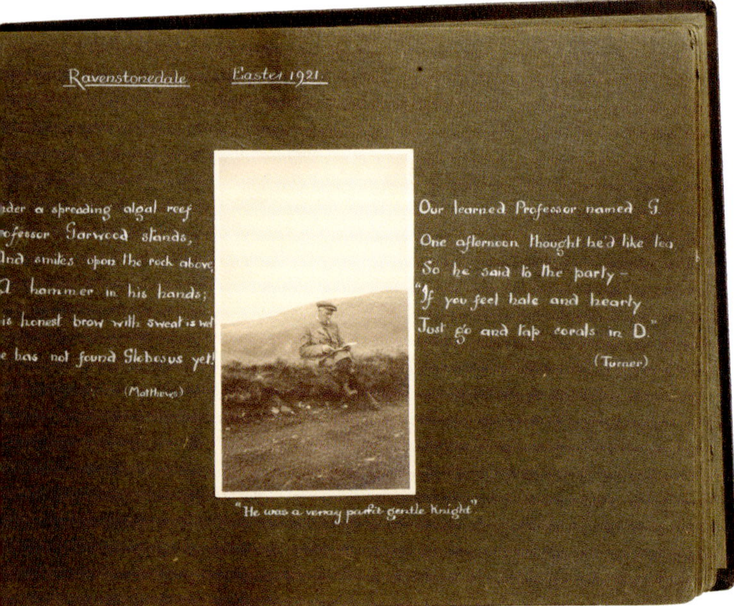

Greenough members' book
1904
Paper and card, width when closed 30 cm
UCL Geology Collections

When students and academics spend time together in the field, a collegial community is formed, and this community is reflected in the notes, photographs and sketches in the albums of the Greenough Society. University life is about so much more than study: it is a chance to forge an identity, pursue your passions, and bond with stimulating, entertaining and intellectually challenging people.

Looking through the collected pictures of the early Greenough Society, their time seems to be so very different to ours, yet a great sense of similarity shines through. Like today's students, they embraced the opportunities to form friendships, to have fun, and to explore the world around them. They also captured and shared their experiences, in a way that is very familiar to modern students. There is a real sense of life, camaraderie and spirit captured in the album, bringing the society's members to life and giving one a glimpse into the lives of some of the people who worked and studied here at UCL so many years ago.

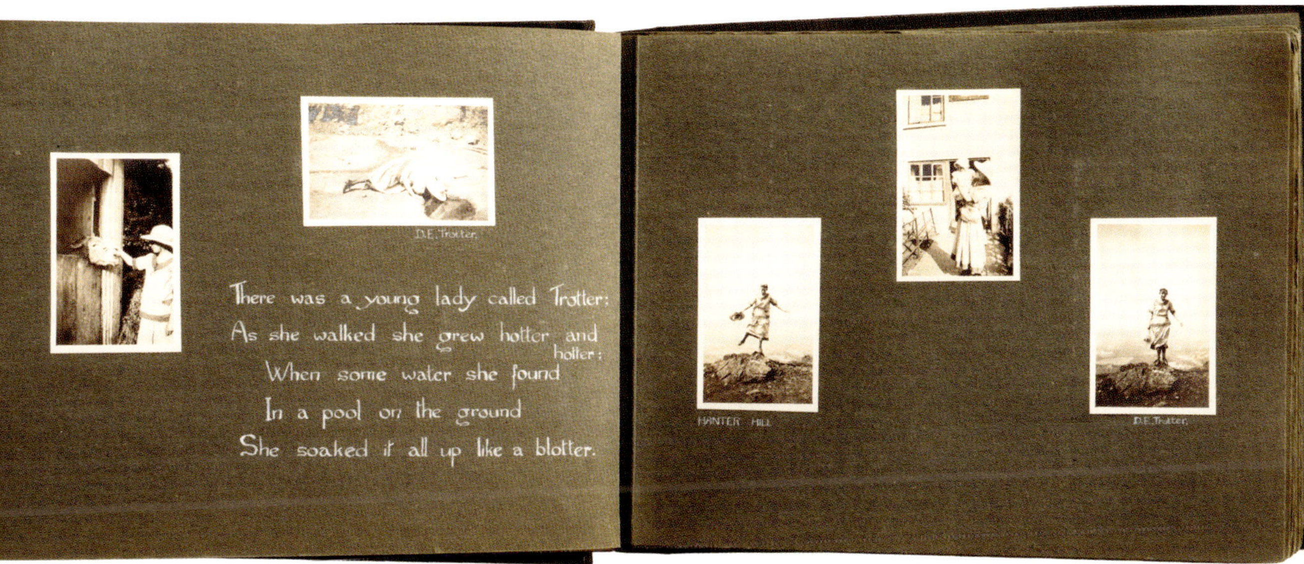

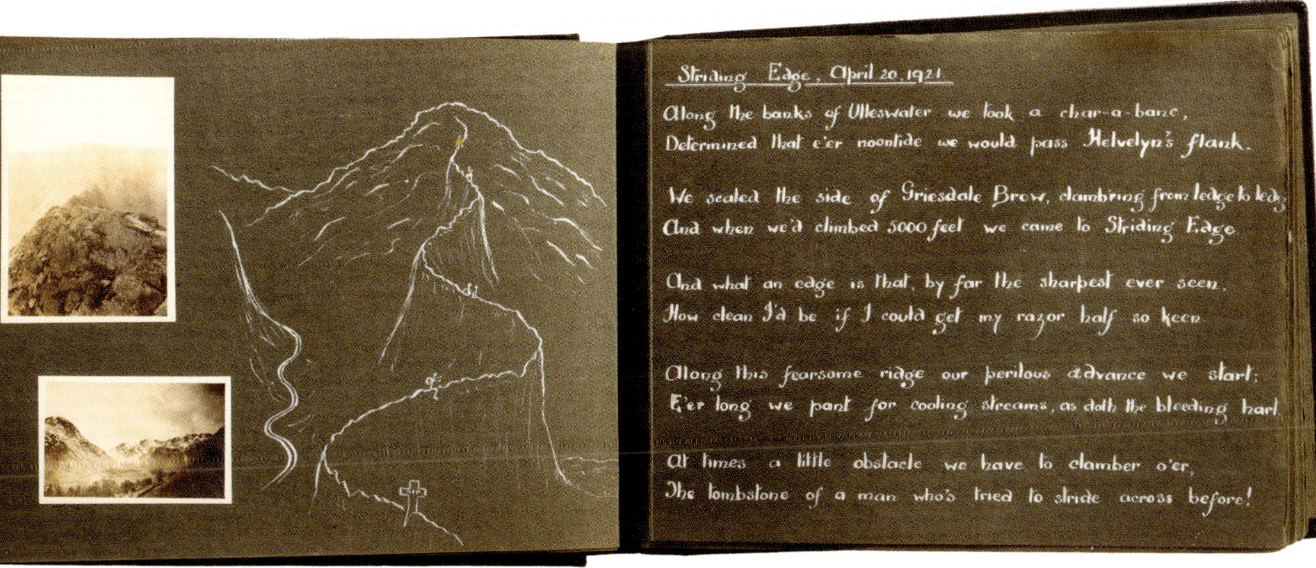

The importance of being blonde

Dr Carole Reeves

Centre for the History of Medicine

THIRTY SAMPLES of synthetic dyed 'hair' in a metal container, manufactured in Germany in around 1905 to the design of Eugen Fischer. Is it a hairdresser's sample box; or a wig-maker's swatch tin; or a killing machine?

To find out let's return to 1905. The German synthetic fibre and dye industries are the finest in the world. Their future products will include versatile materials such as rayon, cellulose and artificial silks, photographic films and chemicals, audio tapes and sticking plasters, agricultural chemicals and life-saving medicines, including the very first germ-killing drug made from a red dye called Prontosil for which its discoverer will win a Nobel Prize.

For the moment, however, no one sees the future. No one opening this flat metal tin and fingering the silky synthetic colourful locks foresees how it too will change the world. For within this container are represented all the races of the world: A to Z – Aryan to Zulu – flaxen blonde to sooty black – smooth, wavy and 'woolly' hair. The numbered hair scale is value-neutral. It is a scientific scale to expose the racially valueless. Hair hierarchy numbered 1 to 30 – Saxon to Semite: Nordic to Negro: Teuton to Turk.

Race science comes of age in the 1920s and 1930s. At UCL, Karl Pearson, professor of Eugenics, uses Fischer's hair scale to examine children of Russian and Polish Jewish immigrants. He concludes they are undesirable aliens, inferior to the native race. At the Kaiser Wilhelm Institute of Anthropology in Berlin, Professor Eugen Fischer examines 400 mixed-race children from the Rhineland and authorizes their sterilization. Their fathers are French colonial soldiers, their mothers Aryan. In cities across Europe and America, bottles of peroxide fly off pharmacy shelves as the importance of being blonde takes on a life or death imperative. The 'bottle blonde' becomes a euphemism for artifice. Being blonde hides the racial status of many of Hollywood's greatest stars – Mae West, Shelley Winters, Kirk Douglas, Paul Newman. Silver, gold and platinum blonde – precious and pure – disguise the America with state laws that forbid marriage between white and black; forbid a child with 'one drop of negro blood' to mix with white children.

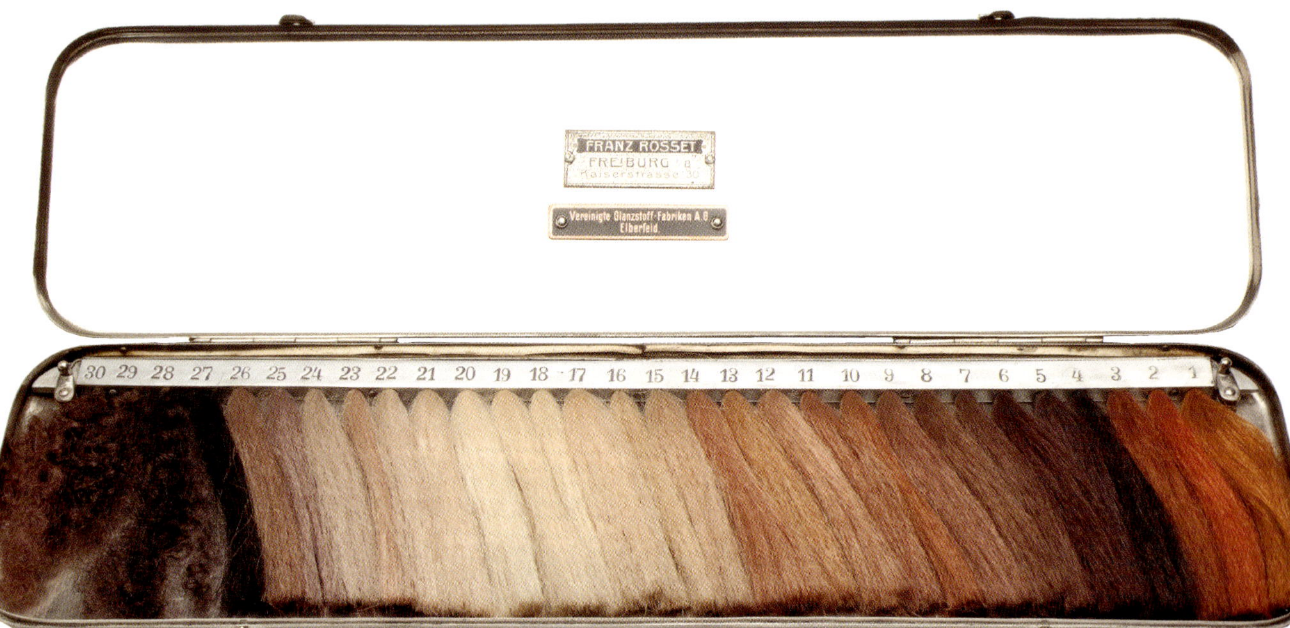

Across Europe during the 1930s the dark-haired inferior races traverse continents to escape the killing machine. Hair samples 27 to 30 – black, wavy, 'woolly'. Wool is the hair of animals. In humans it denotes a status way down the evolutionary scale – up from the ape, but only just. The hair scale is scientific. It is a 'standard' scale which means that all race scientists invest in its truth. The dark-haired races cannot escape the truth. At Auschwitz-Birkenau, Bergen-Belsen, Dachau, Treblinka, Hadamar, hair shaved from those who perish rarely matches samples 12 to 24. Most are piles of clipped ravens' wings.

It is 1941. At Hadamar Euthanasia Centre the ten thousandth cremation is celebrated by staff with bottles of beer – blonde of course.

Haarfarbentafel
Date unknown
Copper-nickel-zinc alloy box, dyed synthetic hair samples 40.5 x 11.5 x 2 cm
UCL Galton Collection
LDUGC-GALT040

The Little Big Man

Professor Michael Worton

Vice-Provost (International)

A FEW YEARS AGO, I started work on a project on the nature and representation of masculinity. This project had grown out of my work on feminism, which explored the cultural and societal challenges for women of living in a society which remains essentially male-dominated. I was fascinated by the fact that some of the most exciting feminist work is done in and through language, whereas there has been little similar experimentation by men in writing about themselves and what masculinity is and means for them. This lack of textual innovation led me to examine the imagery and especially the contemporary photography of the male body by men and I started work on an exhibition, *Typical Men* (2001–02), which aimed to present images which make people think about what we mean when we speak of a 'typical man'.

As part of the preparation for this exhibition, I went back through the history of the representation of the male body. I was struck by how much classical Greek sculpture has influenced some of the most important contemporary photographers, including Robert Mapplethorpe, George Dureau and Arthur Tress. But that is only part of the story. We need to go back further to Archaic Greek sculpture and Ancient Egyptian sculpture, notably the *kouros*, which is a sculptural representation of the male. In these life-size – and sometimes colossal – statues, the men stand rigidly to attention, looking straight ahead, with their arms hanging by their side, fists clenched and feet firmly planted on the ground, though with the left leg slightly in front of the right.

What is striking about these sculptures is their monumentality and the worrying yet mesmerising blankness of their regard as they stare into space. They have broad shoulders, strong buttocks and bulging thighs, and while we may find them amusingly caricatural today, we should remember that similar images of idealised masculinity still fill underwear adverts, body-building magazines and 'action' movies.

The exhibition stressed the importance of recognising that no man is actually 'typical', and that masculinity is just as complex and fluid a category as femininity or woman-ness, and over the past decade, I have continued to explore the different men and masculinities found in paintings, sculptures, novels, and poems, where men are presented as fathers, in sickness, ageing, disabled, insecure as well as happy, hopeful and self-confident. Yet throughout my researches I was haunted by the monumental staring blankness of the *kouros* figures. Then one day in the Petrie Museum I saw this Greek statuette wearing an Egyptian loin cloth, and everything fell into place. He stands in the rigid pose of the *kouros*, but there is in his smallness, in his tensely held arms and his half-open mouth, an intimacy and vulnerability which imitate but also undermine the proclamative dimension of the usual *kouros* figure. Only a few inches tall, he is literally a little big man, but he also thus urges us to think about the paradoxes of masculinities both past and present. Maybe we all are in our own different ways Little Big Men.

Painted limestone Greek statuette of standing male figure
Naukratis, Egypt, Dynasty 26, about 600 BC
Limestone, 23.3 x 6.8 x 5.8 cm
UCL Petrie Museum of Egyptian Archaeology
LDUCE-UC16469

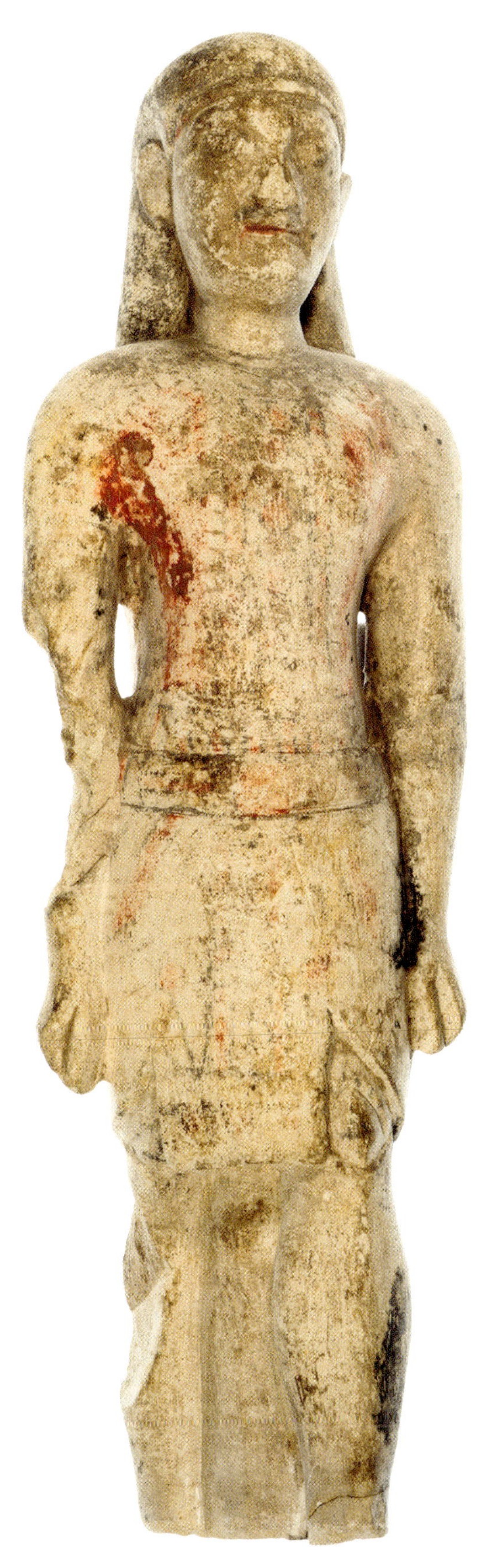

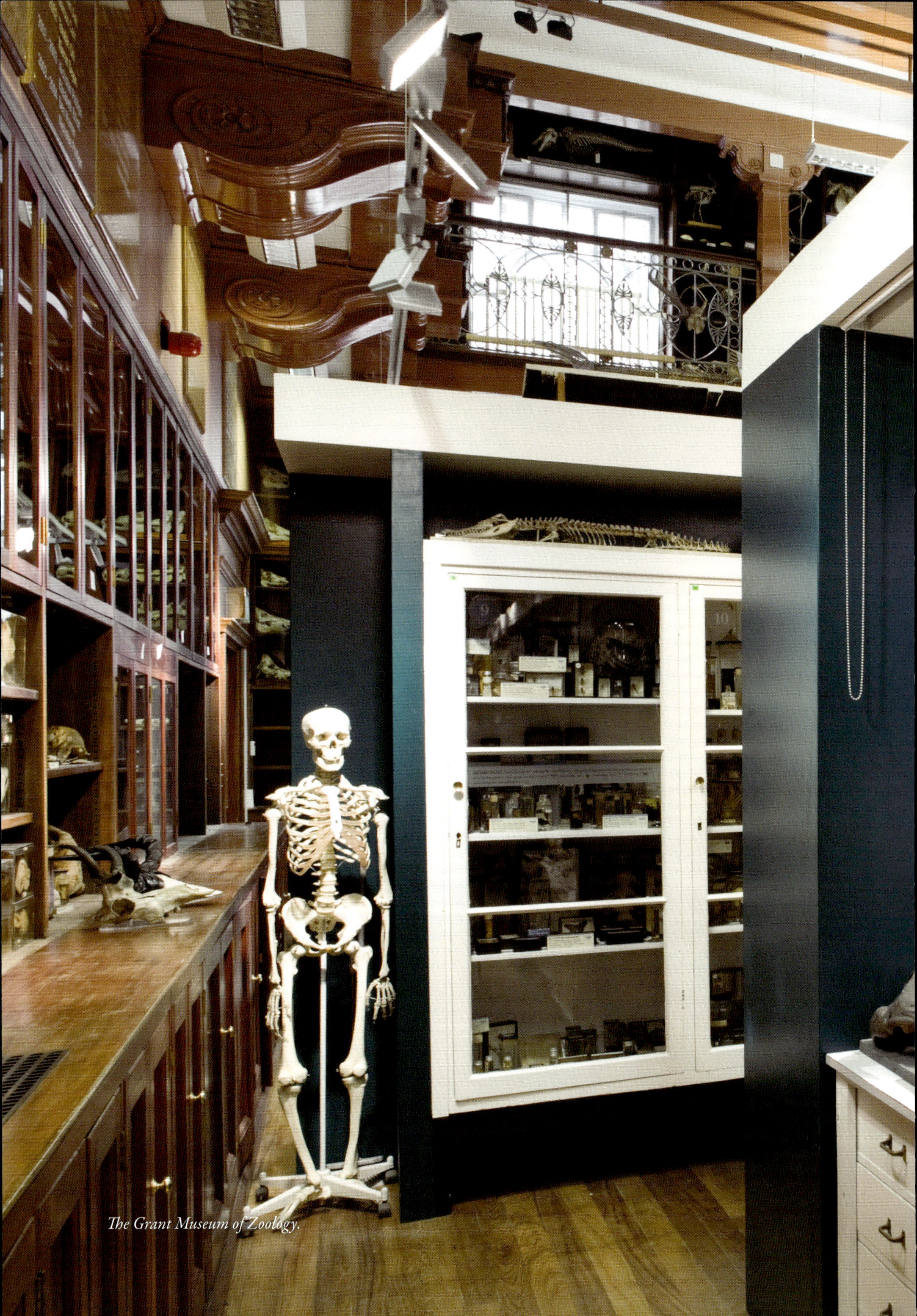

The Grant Museum of Zoology.

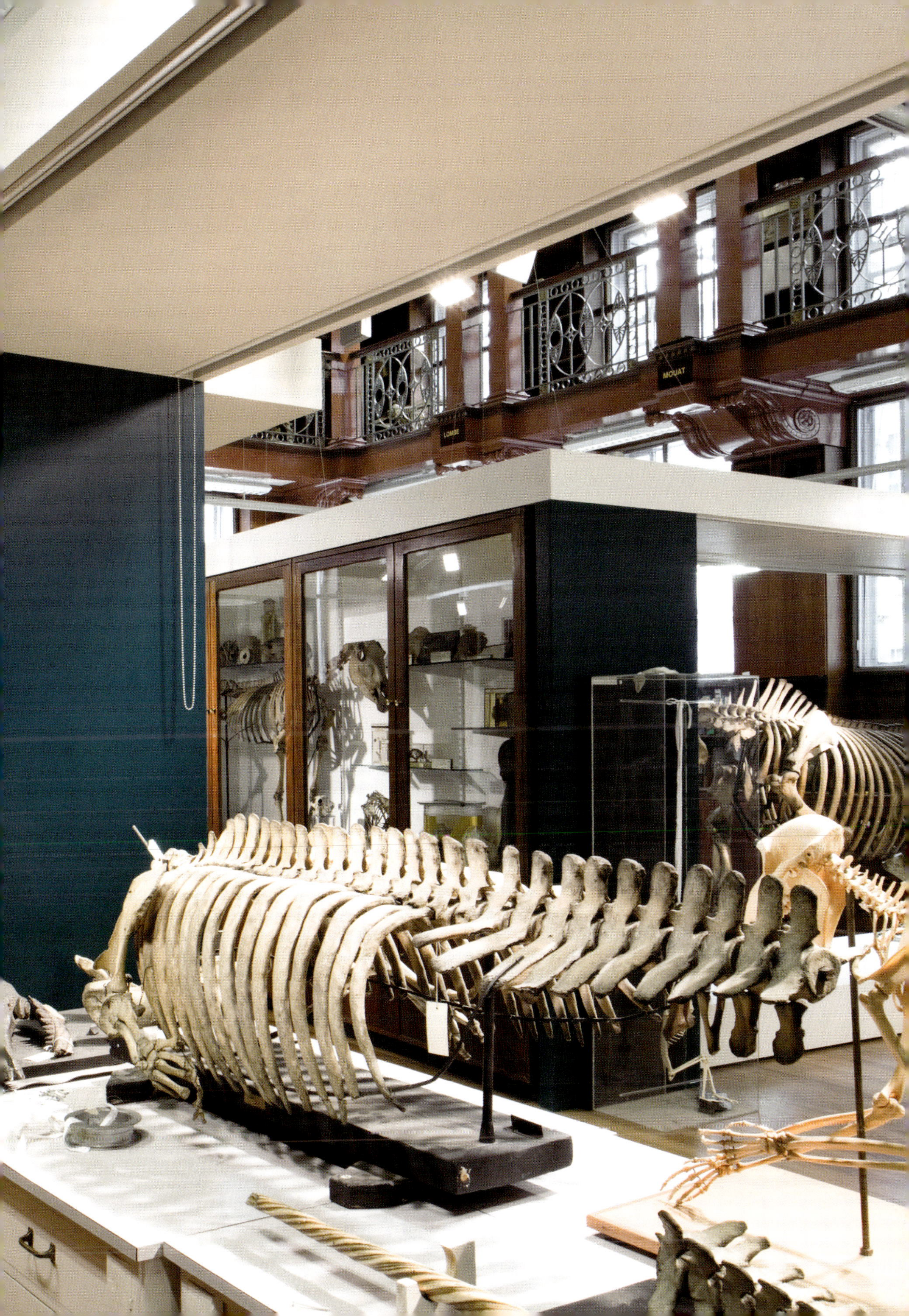
LOMBE
MOUAT

Green Teeth

Professor Susan Collins

The Slade School of Fine Art

BEING ASKED TO CHOOSE an inspiring object presented a nigh on impossible task as the UCL collections are vast, with innumerable objects that fire the imagination in so many directions.

The Grant Museum, as a 'Museum of Museums', has always particularly interested me, as have the objects within it. A part of the collection that drew my attention a number of years ago was the bat skeletons and preserved remains. The fact that they were also available for adoption was a bonus.

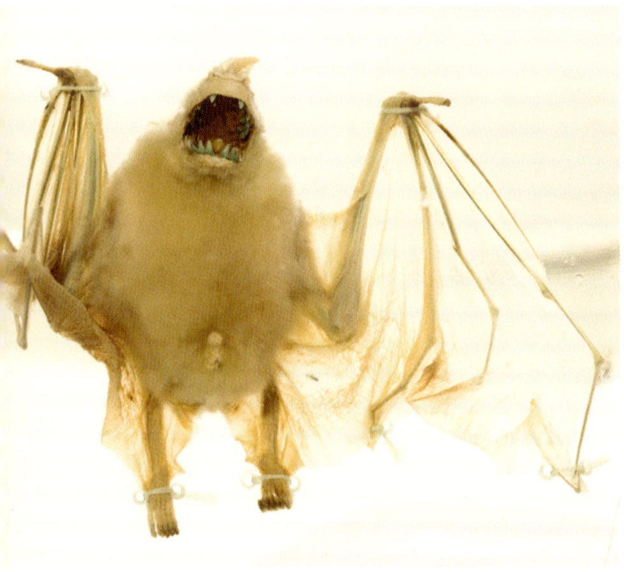

The particular creature I eventually selected was the white-lined bat *Platyrrhinus lineatus*, which is in the same family as the infamous vampire bats. I chose it mainly due to its filmic potential and suggestion of animation. Trapped and shackled in the cramped space of its glass cage, it appears in a suspended asymmetry, not able to fully spread its wings, with its coppered green teeth open *Jaws*-like as if in a silent scream.

Unlike some vampire bats which really do only come out at night (but never under a full moon), and feast only on blood (albeit sleeping quadruped mammals such as horses, cattle and pigs rather than Hammer House of Horror damsels), and despite its fearsome name and appearance, *Platyrrhinus lineatus* is actually a small fruit-, insect- (mainly moths) and nectar-eating vampire bat from South America.

As an artist my work is often made in response to specific sites or situations. I tend to interrogate my subjects, extrapolating a train of thought, or logic, in the process. In this instance it is the gap between the oft-feared fantastical creature of myth and fiction and the reality of its diminutive, captured remains that intrigues me: the image of this fragile creature laid bare for all to see, holding the potential to inspire a whole filmic narrative.

Fluid-preserved bat, Platyrrhinus lineatus
14 x 16 cm
Grant Museum of Zoology and Comparative Anatomy UCL
LDUCZ-Z625

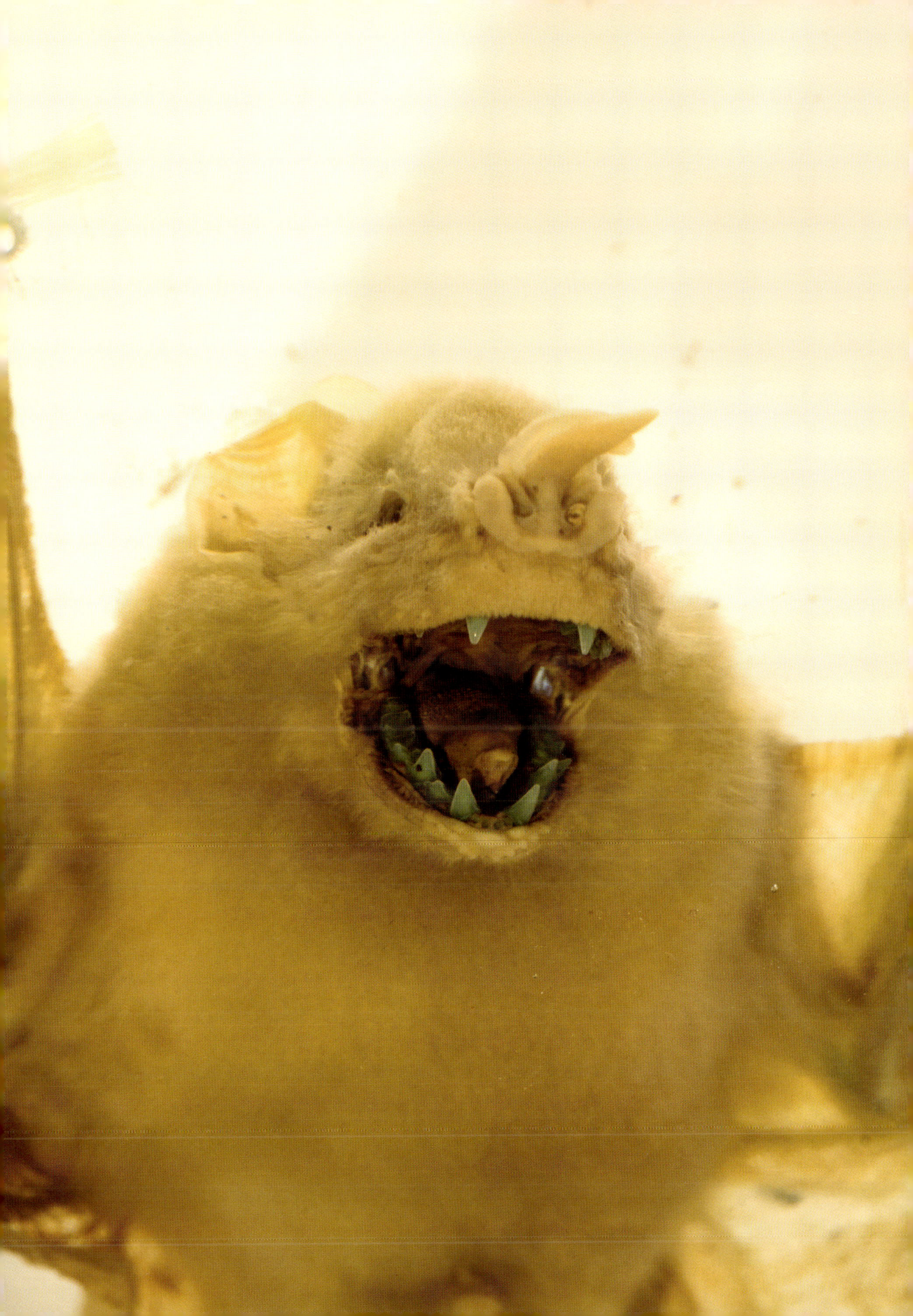

Counting time with peg holes

Professor Sacha Stern

Department of Hebrew & Jewish Studies

THIS PERFORATED AND lightly decorated plaque is not very spectacular; yet it captures, on its own, how daily life might have been organized in Iron-Age Levant. Discovered by the UCL archaeologist Flinders Petrie in Tell Fara in southern Palestine, it is one of several similar artefacts from the area of ancient Judah (others were found in Jerusalem, Lachish, Gezer and Aroer). Carefully perforated with three rows of ten holes, this pocket-size object made of animal bone must have had a meaning, or more likely, a practical function.

Was it a counting device for accountants and other general users? I do not think so, because no one in the Ancient Near East counted up to 30. The Hebrew Bible counts with a base of 10, just as we do today; the ancient Mesopotamians combined the base of 10 with one of 60 – as we count the seconds and minutes of the hour. The 10-hole rows in this plaque suggest that its users counted with a base of 10; but the three rows, totalling 30, call for a more specific interpretation.

As Petrie rightly sensed, a counter with 30 holes was probably designed to track the days of the month. A peg could have been inserted and moved along the holes on a daily basis, so as to keep track of the current date. In the Aroer plaque, there is an additional 12-hole row, for tracking the twelve months of the year. Peg-hole calendars, or *parapegmata* in Greek, are well attested in later antiquity; they are usually inscribed with explicit calendar annotations. This uninscribed device from Iron Age II (early first millennium BC), says something of how time was managed much earlier.

A 30-day, monthly tracker could have served the needs of many different calendars in this early period. In the Egyptian calendar, all months counted 30 days. The same is implicit in some passages of the Hebrew Bible, such as the story of the Flood (Genesis 7:11 and 8:3–4). In Mesopotamia, months were strictly lunar and counted either 29 or 30 days, for which our counter would have been equally suitable; and a similar lunar calendar is likely to have been used in the Levant. Much later in antiquity, in southern Arabia, the month was divided into three *decads* (10-day periods), just as on our plaque; and so was the usage in ancient Greece. Our plaque could easily have been of multinational use.

Having worked for some years on ancient calendars, I was amazed to find this object at UCL, on my very door-step. In my last book, *Calendars in Antiquity* (2012), I argued that in spite of their great diversity, ancient calendars shared a common history. Political factors, such as the expansion of great empires in the first millennium BC, were conducive to the development of calendars in similar directions and to their gradual standardization. Although confined to the area of Judah, this simple plaque, which could have been equally shared by so many calendars, is perhaps a material embodiment of this emerging, common culture in the ancient Near East.

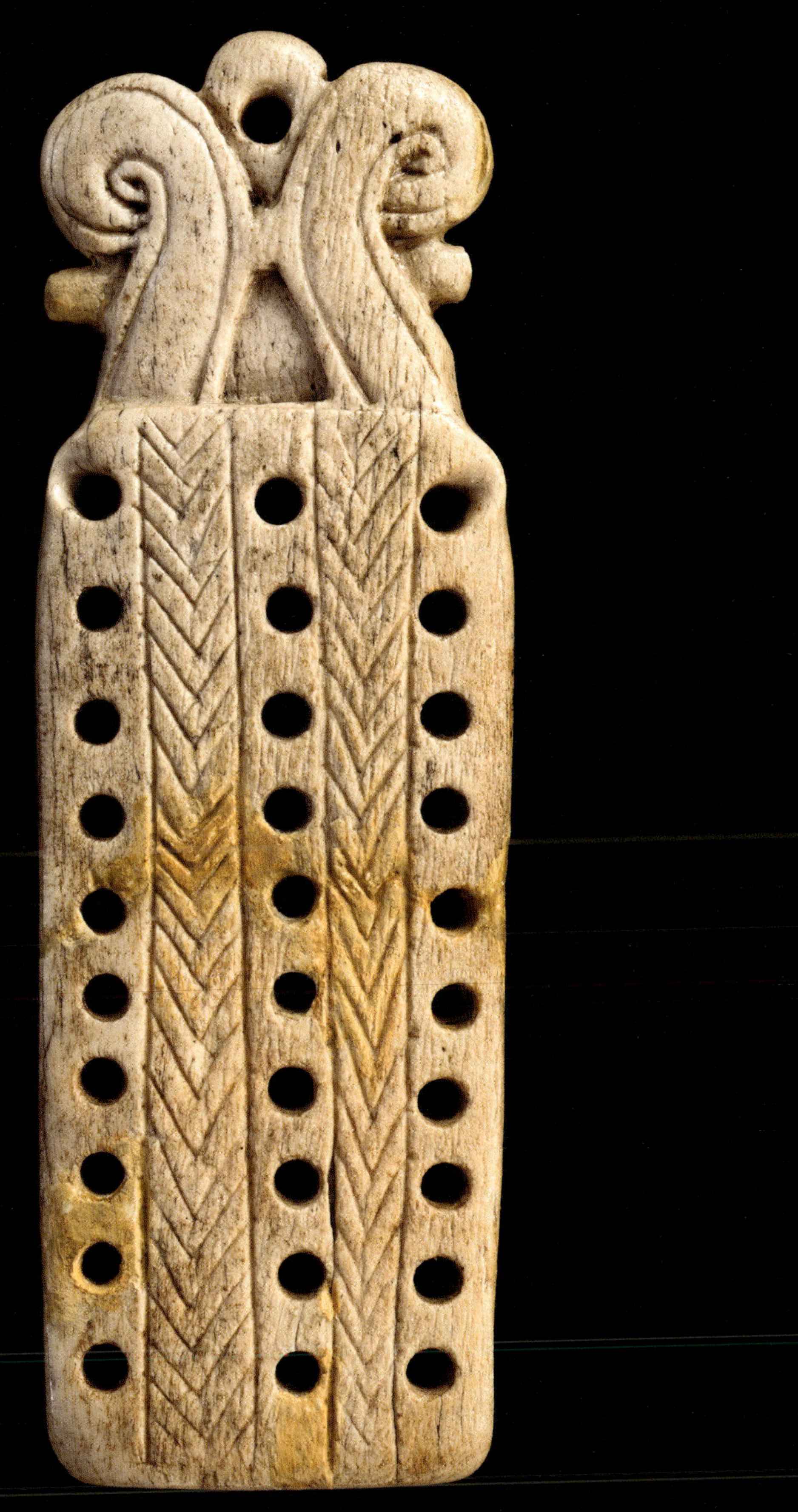

Urban transitions: looking back, looking forward

Professor Caren Levy & Dr Adriana Allen

The Bartlett Development Planning Unit

JUST AS IN CALVINO'S *Invisible Cities*, where Marco Polo explains the cities he has visited to Emperor Kublai Khan through objects from each site, the cereal grains from Abu Hureyra tell us about an urban transition that started 12,000 years ago in the Fertile Crescent of the Euphrates.

The grains from Abu Hureyra talk about the domestication of seeds and agricultural cultivation and how the production of surplus food yields led ultimately to the emergence of urban settlements. While this complex process of interaction between human beings and their environment took some 5,000 years to evolve, the rate of global urbanization in contemporary times has changed dramatically in only 150 years. Between 1850 and 1950 urban dwellers grew from approximately 4 per cent to 25 per cent of the world population. This went up to 40 per cent in 1990, 50 per cent in 2008, and is estimated to rise to 75 per cent in 2050, indicating the unprecedented contemporary and future urban change with which citizens, politicians, planners and policy-makers are grappling. The highest proportion of this change will be in the urban centres of Africa, Asia and the Middle East, with continued urban growth in Latin America and North America, and marginal growth in Europe. The associated changes in social organization and economic development, along with the production and transmission of knowledge and technology, are daunting in their speed and tendency to multiply and deepen inequalities and conflicts, to degrade the environment and to reshape the built environment. These changes are also inspiring in their creation of prosperity, alongside the production and dissemination of knowledge and technology of all kinds, ranging from the capacity to address disease to the making of cities.

Abu Hureyra has 'disappeared' twice in its long history, both times for reasons that have resonance for our times. The settlement was abandoned in about 7000 BC because of soil degradation (through over-cultivation and grazing) and climate change. Ironically, the impetus for the excavation of the site between 1971 and 1973 was the building of a dam by the Syrian Arab Republic on the Euphrates River at Tabqa, now known as Medinat el Thawra. In 1974, the site, along with the surrounding area, was flooded. Given the speed in the growth of our cities and the enormity and diversity of their ecological and social footprints, this story is cause for sober reflection on the future of our planet. Are we able to imagine a sustainable and just present and future under a scenario of almost total urbanization?

Archaeology in sites like Abu Hureyra reminds us that, as with the multiple global origins of the domestication of seeds for cultivation, there are multiple paths to the future – but all requiring food, water and related resources. The diversity and creativity of communities to think and act 'glocally' while embracing democratic and inclusive principles, are the strongest hope to find alternative ways of planning and managing our shared space in an urban future.

Grain samples
Abu Hureyra, Neolithic 8000–6000 BC
UCL Institute of Archaeology Collections

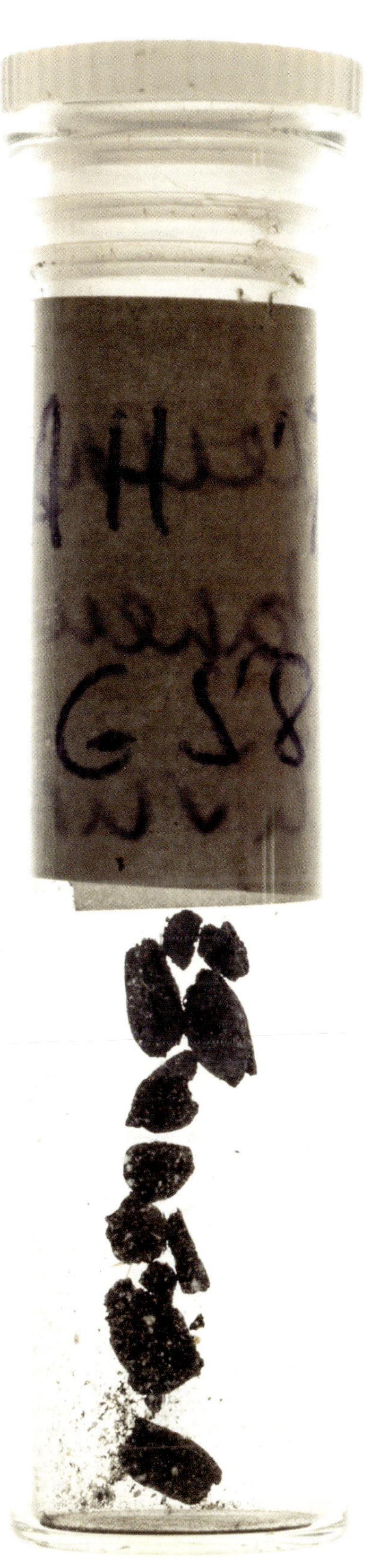

Childe's notebooks

Professor Stephen Shennan

Institute of Archaeology

P ROFESSOR VERE GORDON CHILDE (1892–1957) was the 20th century's leading prehistorian of Europe and a leading theoretician of archaeology. He was also director of the Institute of Archaeology from 1946 to 1957, the position I now hold, and my specialist field in archaeology is essentially the same as his, so there is something very special for me in looking at the notebooks of his in the institute's collections.

Childe was an Australian who studied at Oxford during the First World War, before returning to Australia and becoming engaged in left-wing politics – he was a passionately committed socialist throughout his life and this very much influenced his archaeology. He subsequently returned to Britain and developed an academic career, becoming the first professor of Archaeology at the University of Edinburgh in 1927, a position he held until he moved to London.

Childe made his name with his first archaeology book, *The Dawn of European Civilisation*, which appeared in 1925, and he continued producing revised editions until the sixth and last in 1957. It represented the first attempt to produce a narrative of European prehistory, based on the only source available, the archaeological finds. Childe was able to do this because he devoted himself to acquiring an encyclopaedic knowledge of that material

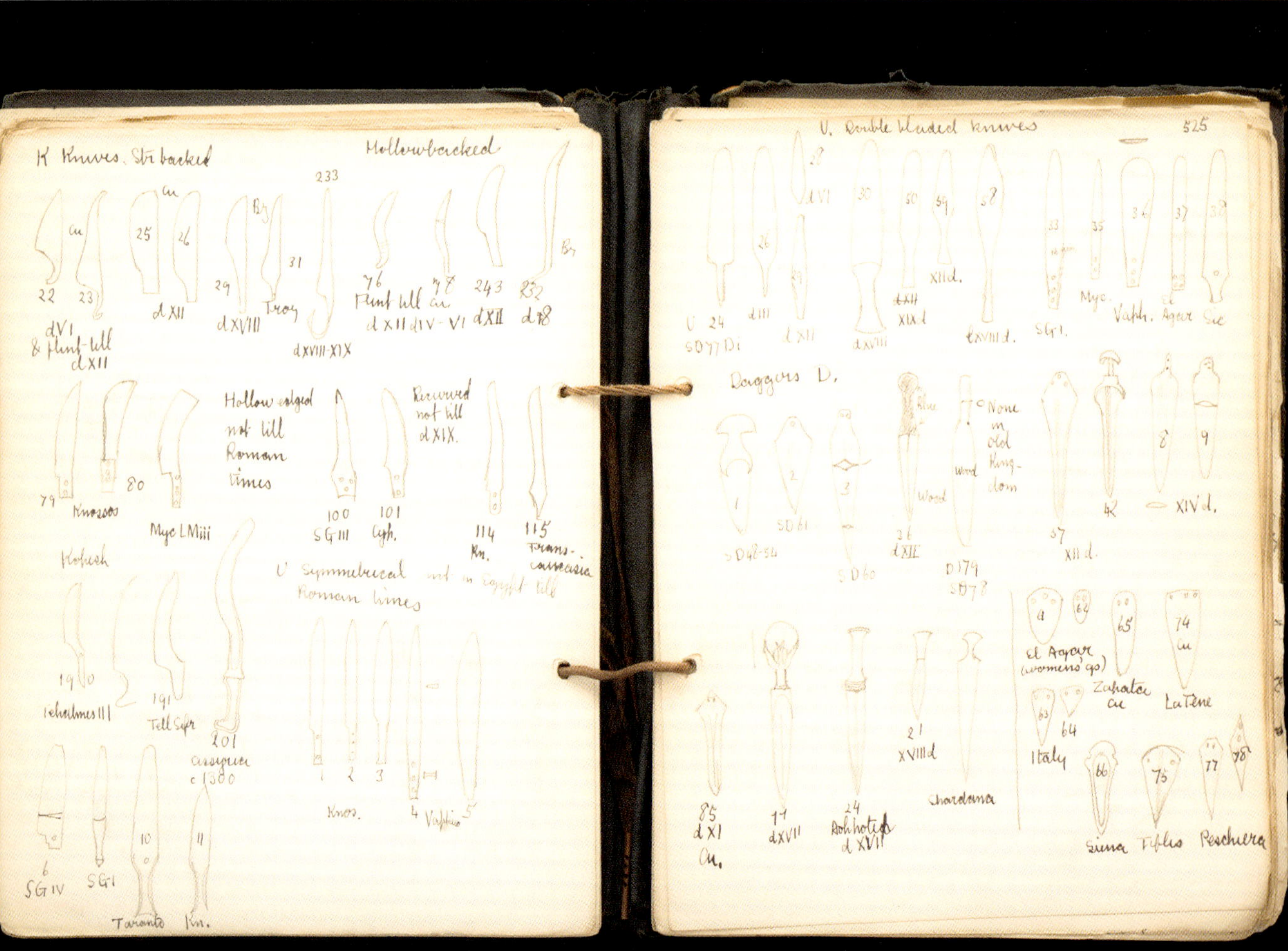

by reading foreign journals and travelling widely to visit museums and colleagues across Europe, including the then Soviet Union. This was unique at the time and even today the number of prehistoric archaeologists who take an interest in countries beyond their own is relatively limited. The notebooks contain the raw material for his syntheses: page after page of neat sketches of pottery, decorated and undecorated, and of copper and bronze weapons, tools and ornaments of all sorts, noted by their find spot. With these at hand Childe could identify and characterize changing regional patterns in artefact assemblages and the similarities between one region and another which suggested patterns of contact and influence. He concluded that the main driving force of European prehistory was 'the irradiation of European barbarism by Oriental civilisation'.

But his comparisons and groupings of material also took Childe in another, more theoretical, direction. In *The Danube in Prehistory* (1929) he developed and extended the concept of the 'archaeological culture', which had initially been formulated by German scholars:

We find certain types of remains – pots, implements, ornaments, burial rites, house forms – constantly recurring together. Such a complex of regularly associated traits we shall term a 'cultural group' or just a 'culture'. We assume that such a complex is the material expression of what today would be called a people.

This concept has probably been the most influential one in all prehistoric archaeology. It continues to be debated to this day and to be used as a framework even by those who disagree with it. In one way or another it remains indispensable.

To see the raw material of these constructions in Childe's neat sketches is a fascinating experience, all the more so to one of his successors as director of the institute feeling the burden of responsibility to the immortal ancestors.

Gordon Childe notebooks
London, 1920–30
UCL Institute of Archaeology Collections

More than just an enigmatic face

Professor Stuart Robson

Department of Civil, Environmental & Geomatic Engineering

THE CATALOGUE ENTRY for UC39694 – 'Limestone shabti of Wsr-h3t. Mummiform with striated lappet wig. Pectoral slung from neck. Hoes held and brick moulds over both shoulders. Four lines incised hieroglyphs preserved and one column behind. Lower half of figure lost and back head chipped' – might be scientifically correct, but it hardly does justice to a lovingly crafted servant for the afterlife. Very few are privileged enough to have hands-on access to appreciate the irreplaceable, but the whole community should somehow have a level of close access that supports detailed inquiry and examination of what our ancestors were capable of achieving. What if we could produce and bring into your home a 3D colour digital copy that retained all of the subtlety of the original work – a digital surrogate of the original?

In 2006 our international search for a 3D colour capture system that could accurately record the surfaces of a wide range of museum objects was completed and we took delivery of an Arius3D colour laser scanning system. This state-of-the-art equipment was unique in Europe and supported a joint venture between UCL and Canadian company Arius3D. The result was 3D-Encounters, a group of highly skilled and talented staff coupled with the specialist computing software and hardware needed to develop, explore and share colour 3D content collected from UCL's Petrie Museum. Since that time, 3D-Encounters has been at the heart of a series of successful digital 3D colour surface projects covering Humanities and Engineering. From its interdisciplinary base spanning the Petrie Museum and the Department of Civil, Environmental & Geomatic Engineering, the team have developed a working best practice for digitally recording museum objects. Whilst not yet having an army of digital shabtis working away in the background, the developed processes guarantee the quality of the information collected and ensure the long-term preservation of that data for future use. Projects have extended well beyond UCL, involving all of the key London museums and reaching out through the exhibition of digital objects as far away as Qatar, Alaska and the Solomon Islands. Strengths taken from the same toolbox of technology have extended the work to cutting-edge aeronautical and space engineering with Airbus and NASA along with an award for best 'small and medium enterprise collaboration'.

So why UC39694? Shabti Wsr-h3t was our first digitization guest. His enigmatic face and challenging incised stone surface provided our first training dataset and the basis for a 3D digital ambassador that became the opening logo for UCL colour-scanning workshops and conferences. These events have brought in specialists and interested individuals from all over the globe. I often contemplate what his unknown maker would have thought had they realised that their afterlife funerary servant would take on a new, seemingly magical, digital form for a key role some 3,500 years later. As we explore the 21st-century ways of seeing and experiencing the world around us, UC39694 has opened for us a new era of digital surrogate quality 3D photography.

Shabti of Wsr-h3t
Dynasty 18, about 1400 BC
Limestone, 18 x 8.4 cm
UCL Petrie Museum of Egyptian Archaeology
LDUCE-UC39694

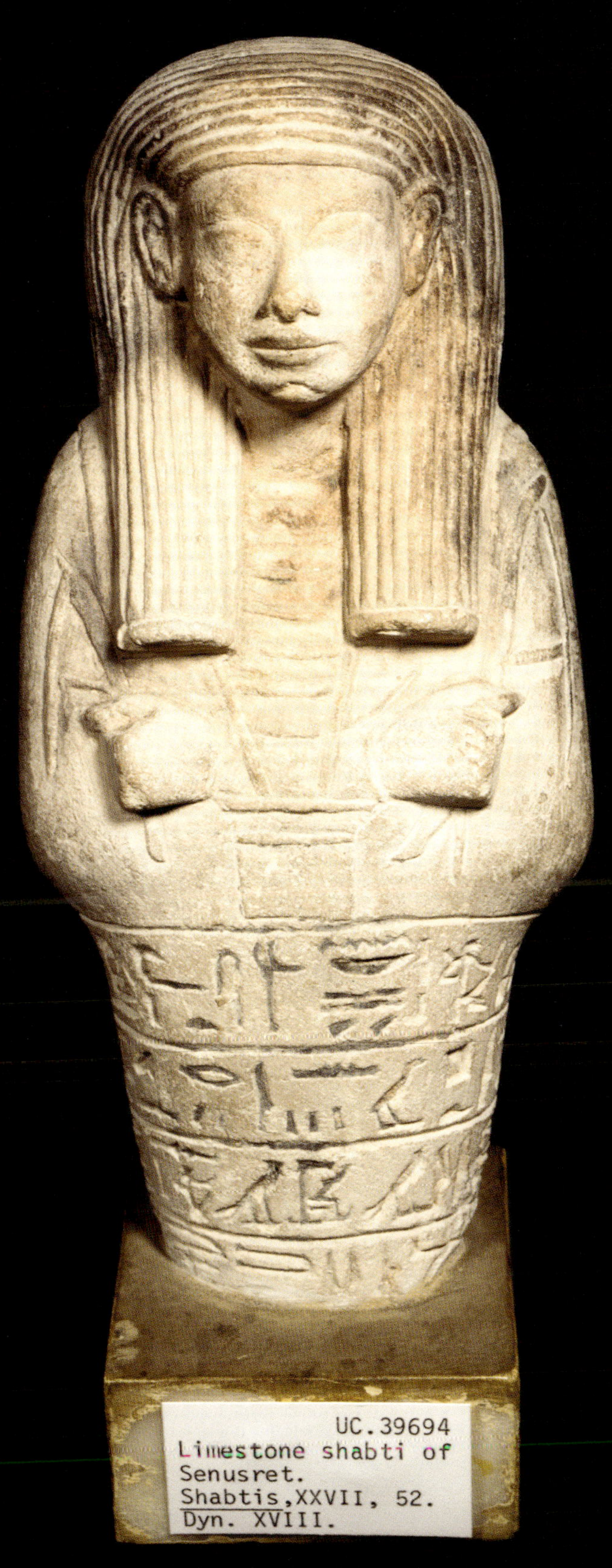

UC.39694
Limestone shabti of
Senusret.
Shabtis,XXVII, 52.
Dyn. XVIII.

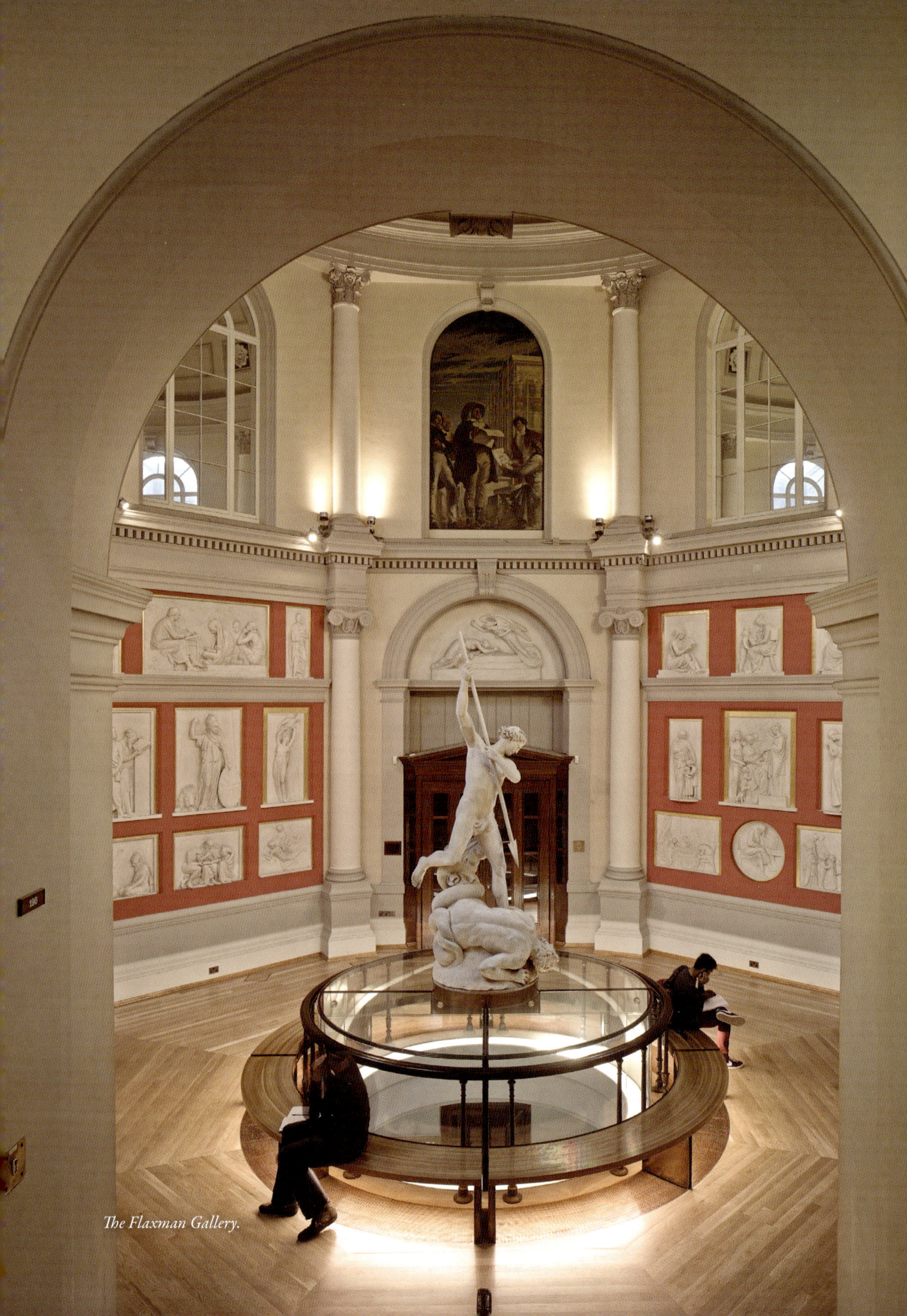

The Flaxman Gallery.

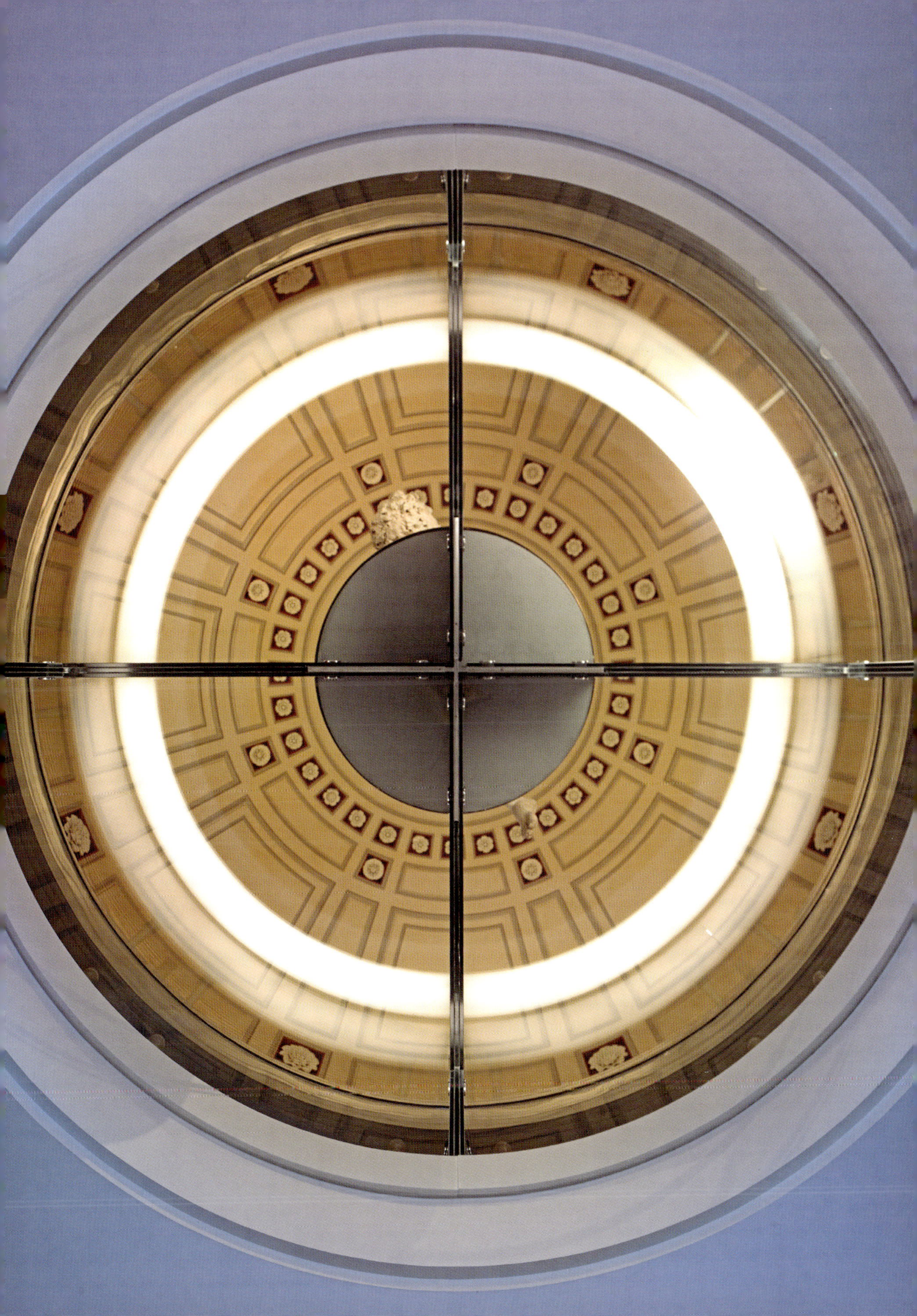

Fleming's Hand Rule

Dr John Mitchell

Department of Electronic & Electrical Engineering

ALTHOUGH JOHN AMBROSE FLEMING'S invention of the thermionic valve, the first electronic device, transformed the world in which we live, it is for a visual mnemonic that he is best remembered. Across the world, generations of physics students have sat in exam halls, holding out their hands stiffly in the familiar shape, muttering the incantation 'Fore finger, Force; seCond finger, Current; thuMb equals Motion'. Traditionally, this is coupled with trying to remember which hand represents motors and which hand represents generators (answer below).

The glass slide shown here is one of over 3,000 glass 'lantern slides' produced by Fleming and his successors and held in the UCL Electronic and Electrical Engineering Collections. The slides illustrate subjects ranging from electrical power distribution and radio communications to astronomy and Fleming's travels in the Middle East. The collection includes many hand-drawn slides, and given Fleming's reputation as an amateur artist it is highly likely that he drew the Right-Hand Rule slide himself. He was passionate about the communication of science and gave a large number of public lectures. Attendees reported that his lectures were clear and well thought out, often with practical demonstrations – although they sometimes complained that he tended to speak too fast.

The chance to study and teach in the department founded by Fleming is a great privilege. It is hard today to imagine one individual making as many contributions, over such a wide variety of fields, as Fleming contributed to electrical engineering during his forty years leading the department at UCL. Most famously, his invention in 1904 of the thermionic valve was the first step in controlling electricity and heralded the birth of electronics. Like many discoveries, it grew from a spark of inspiration: in this case the memory of the experiments he had conducted nearly twenty years before into why contemporary lightbulbs darkened during use. After some searching, an old bulb was retrieved from a store cupboard and put to work in a completely new role as a component in the processing of wireless signals. His new device subsequently turned out to have many other uses and sparked the development of the electronics revolution.

Although many alternate labels have been suggested to accompany the characteristic gesture, the essence of Fleming's mnemonic has lasted for over a century and is still going strong. And although I doubt many students could tell you anything else about Fleming, the power of this aide-mémoire is demonstrated by the fact that it can still be recalled by almost all who have met it. It just goes to show the lasting impact of a good teacher.

Answer: Fleming's left-hand rule is for motors and Fleming's right-hand rule is for generators.

Right-hand rule glass lantern slide
Date unknown
Glass, 8.3 x 8.3 cm
UCL Electronic and Electrical Engineering Collection
LDUSC-CC465

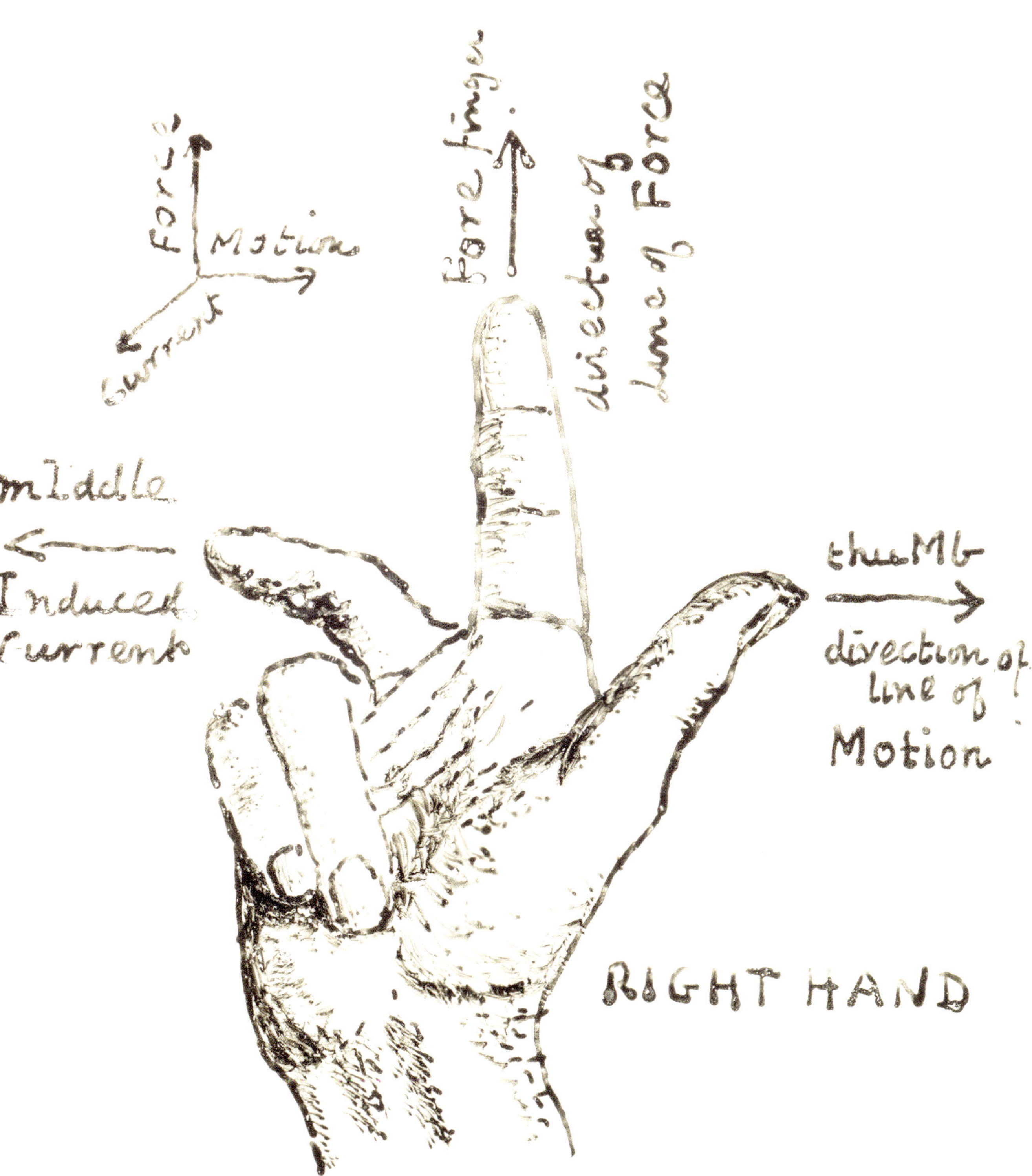

Force
Motion
Current
Fore finger
direction of
Line of Force
middle
Induced Current
thuMb
direction of
line of !
Motion
RIGHT HAND

Tweeting Sheep Pig

Dr Andrew Hudson-Smith

The Bartlett Centre for Advanced Spatial Analysis

SHEEP PIG IS SO CALLED as he looks like a cross between a sheep and a pig. Crafted out of clay and purchased from a branch of the home furnishing store Cargo, the anonymous-looking office doorstop is actually famous for being one of the first objects connected to the internet and able to 'store' and 'recall' memories of itself.

The doorstop is part of a research project called 'Tales of Things and electronic Memory' (TOTeM) exploring social memory in the emerging culture of the Internet of Things. It sounds complicated but it is really a mix of the concepts behind eBay, Facebook and the Antiques Roadshow, so every object has its own web page and is able to recall its past via its own timeline. Sheep Pig is wired up with a series of radio-frequency identification (RFID) tags and black and white printed labels known as QRCodes. If he is moved near a scanner or smart phone he is able to tell the story of his past and capture the present. He is able to recall information in text, video or audio form, complete with a geo-location, details of owner and date: in short he is able to recall his history.

If you scan one of the tags on Sheep Pig he will tell you that he used to be part of a pair of doorstops. His other half was a cow who sadly met her fate in a domestic accident. This accident led to Sheep Pig being brought into the offices of The Bartlett Centre for Advanced Spatial Analysis (CASA) at UCL. Sitting in the office he was a natural choice to become not only the first connected object of 'Tales of Things', but also the first object that was able to tweet to the outside world, securing his place in the history of the Internet of Things.

TOTeM allows objects to record and replay memories of their time. It is in some ways an internet of second-hand goods and in terms of research has implications upon production and consumption to transform the way in which people shop, store and share products. Every object has its own story, web page and Twitter account, opening up the potential to transform how we treat objects, care about their origin and use them to find other connected objects.

You can connect your own objects to the Internet of Things via http://www.talesofthings.com

Sheep Pig will carry on being an experimental object in the realm of the Internet of Things, and the next step for researchers at CASA is exploring the concept of objects transmitting emotions and empathic feelings via the network.

Sheep Pig doorstop
Purchased in London, date unknown
Clay, diameter 18 cm
UCL Centre for Advanced Spacial Analysis

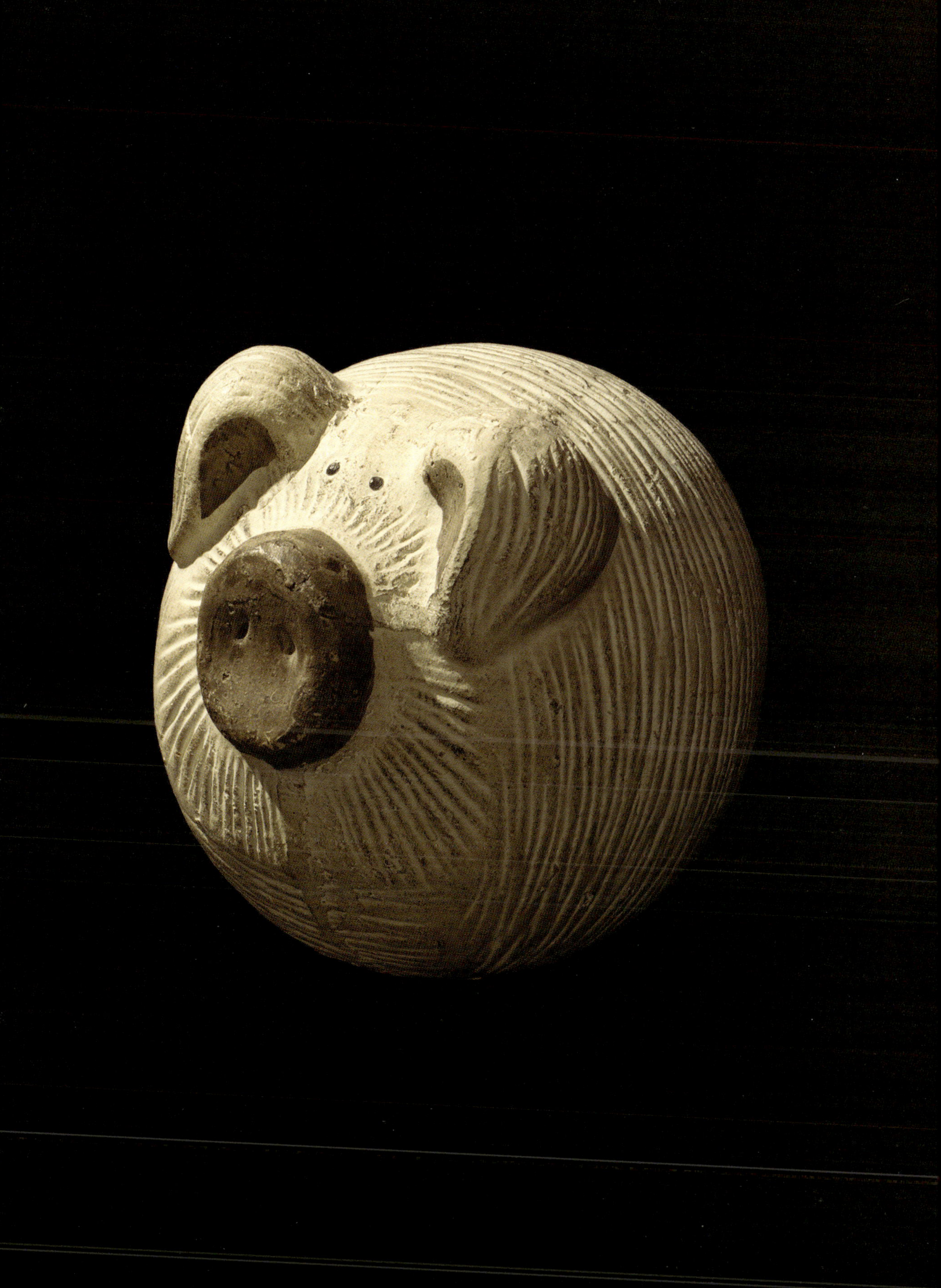

The Pares Silver

Dr Robin Aizlewood

School of Slavonic and East European Studies

O N 21 JUNE 1909, a delegation from the Third Russian Duma began a sixteen-day visit to Britain. The visit had been organized by Sir Bernard Pares, professor of Russian History at Liverpool University, who had travelled extensively in Russia and was well acquainted with members of the Duma.

A letter of invitation to the Duma was signed by the archbishops of Canterbury and York, the duke of Northumberland, the earls of Cromer and Jersey, Lords Avebury, Brassey, Curzon, Rayleigh, Reay, Sanderson and Weardale, the speaker of the House of Commons, the lord mayors of London and Liverpool, parliamentarians, academics and editors of the British press.

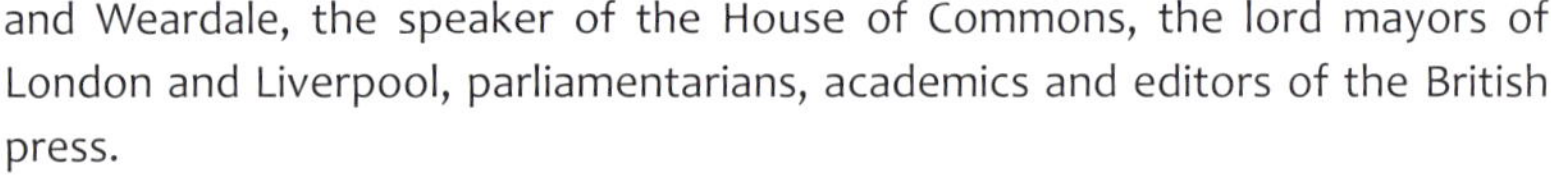

The visit was intended to 'mark a feeling of cordial good-will between the two countries and the desire on both sides for closer relations, intellectual, political, social and commercial'. The major part of the visit was spent in London, where the delegation was received at Buckingham Palace and at the Houses of Parliament. The delegation also spent one day each in Edinburgh, Liverpool, Oxford and Cambridge.

This was the Duma's first foreign visit and there was a great deal of excitement about it in Britain, with extensive press coverage. After the visit Bernard Pares acted as secretary to a committee headed by Lord Weardale to take action whenever any prominent Russian visited England and facilitate meetings according to their interests, be they ecclesiastical, social, literary or commercial.

Shortly after the visit Sir Bernard Pares received, as he described in his memoirs, 'a magnificent silver punch-bowl and salver' with little goblets, studded with stones, each bearing the name of a member of the party.

There was a reciprocal visit by British parliamentarians in 1912 headed by Lord Ullswater with Sir Bernard as secretary and including Lord Weardale and five other members of the House of Lords, ten from the House of Commons, four bishops, three generals and Lord Charles Beresford representing the Navy.

In 1922 Sir Bernard Pares became the director of the School of Slavonic and East European Studies (SSEES), a post he held until his retirement in July 1939. The silver was donated to SSEES in 1975 by the Pares family, and is on permanent display in the foyer of the SSEES building.

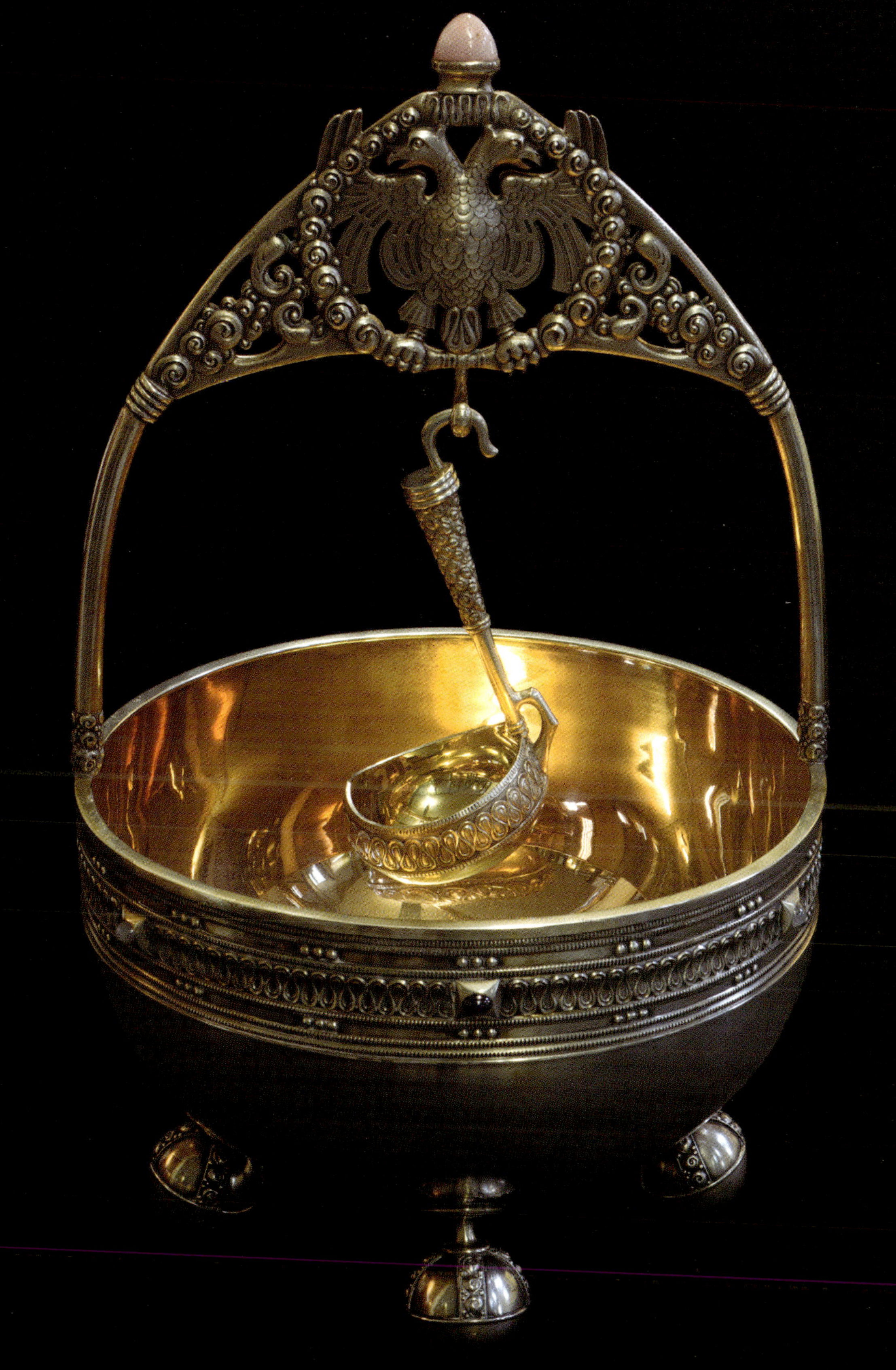

Phoenician glass pendant

Dr Rachael Thyrza Sparks

Institute of Archaeology

THIS TINY GLASS PENDANT takes the form of a bearded man, staring enigmatically out from a distant past. It started its life as a disparate collection of raw silica and colourants, but was then crafted at high temperatures into this miniature personality. It's not a portrait, but the hairstyle and magnificent beard would seem familiar to its makers, mirroring the faces to be seen strolling around the streets and harbours of their Phoenician homeland.

At some point, this object made its way from a Phoenician workshop to a small city in the southern Levant, modern-day Tell Jemmeh. Did the person who wore it care about its origins; did they value its exotic character, or did they just like the way it looked? Either way, imagine how its owner felt when the necklace bearing this fragile pendant snapped and the beads scattered over the ground. Where did it fall and why didn't they pick it up? It was found in what was probably a fill layer beneath a Persian-period house, so the soil it lay in was probably swept up and brought there from somewhere else in the city. Maybe it was lost in a crowded marketplace, and in scrabbling to collect the pieces the owner missed this tiny object as it rolled under a stall, between the feet of a lounging Persian soldier or into a pile of donkey droppings.

Such an attractive object was bound to be cherished and loved by its wearer, but what if it was more than just a pretty bead? What if it was a lucky charm, something to keep adversity at bay? Losing one's luck must seem like the gods have turned away from you, leaving the worry of bad fortune to come.

More intriguing perhaps is the question of how such a delicate object remained intact. Made of glass, it could easily have been crushed by a misplaced foot or fallen mudbrick. Yet it somehow went into the archaeological record unscathed and came out safely the other side – an object that clearly laughed at the perils of the workman's pick or the short but deadly flight from a basket of spoil into a harsh metal sieve. Did a ray of light catch the shiny glass surface and cry out at the excavator from its dull brown surrounds? And how did the finder react? One thing seems likely: whoever found it would have profited, as the excavator of Tell Jemmeh, Flinders Petrie, paid a special finder's fee, known as *bakshish*, for prime discoveries of this kind.

The value of our bearded man may have once rested in questions of cost – of manufacture, purchase, loss and discovery. Its value today is somewhat less tangible: a face that takes us back in time to unfamiliar lives in an unfamiliar landscape, linking us to an otherwise unreachable past. This object may be tiny, perhaps, but not insignificant.

Phoenician head pendant
Tell Jemmeh, 500–300 BC
Length 2.5 cm
UCL Institute of Archaeology Collections
LDUAC-EXXXVI.25/35

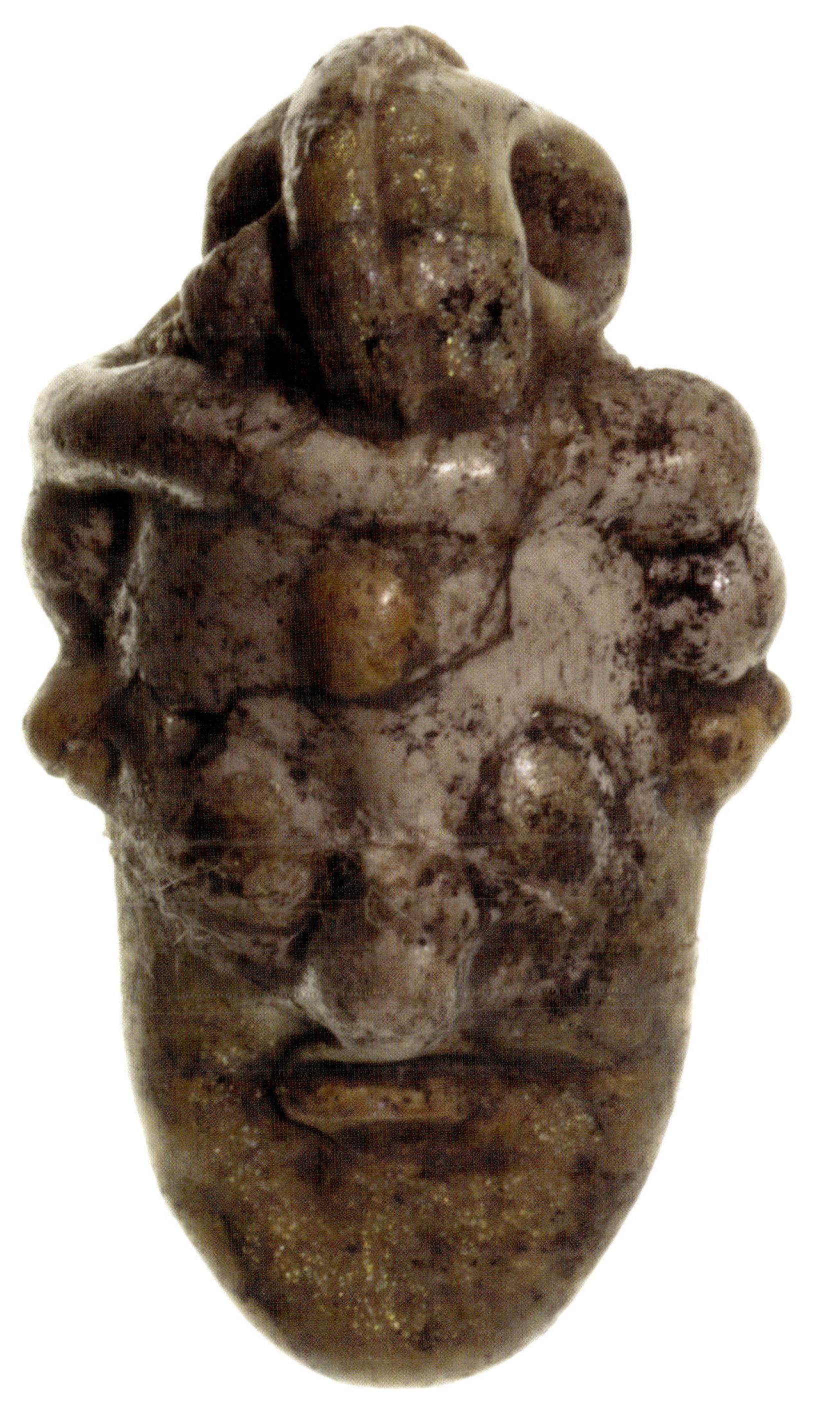

Bentham himself

Professor Paul F. Snowdon

Department of Philosophy

O F ALL THE INTERESTING OBJECTS in UCL that have from time to time inspired me, the one that has most exercised this role is Bentham's so-called 'auto-icon'. Whenever I come into UCL I enter via the South Cloisters, where it is housed, and glance at it. It is not visually attractive or engaging and no one would think that the object in the box actually is Bentham, since it is simply Bentham's skeleton, clothed in Bentham's clothes, and with a wax head resembling, apparently to a remarkable degree, the head of Bentham. But to see this, knowing of its link to Bentham, is inspiring because Bentham was one of the great social thinkers and reformers of the early 19th century. Bentham's project was to think through and implement the utilitarian approach to evaluating action and social institutions. The central idea is that actions and social organizations are not good or acceptable unless they maximize the happiness of the human beings they affect. Bentham, who had been an amazing child prodigy, devoted his life to thinking this fundamental vision through in a more complete way than anyone else. In the course of doing so he developed many views which are remarkably ahead of his time. Thus, he criticised the treatment of women in society, and the cruel treatment

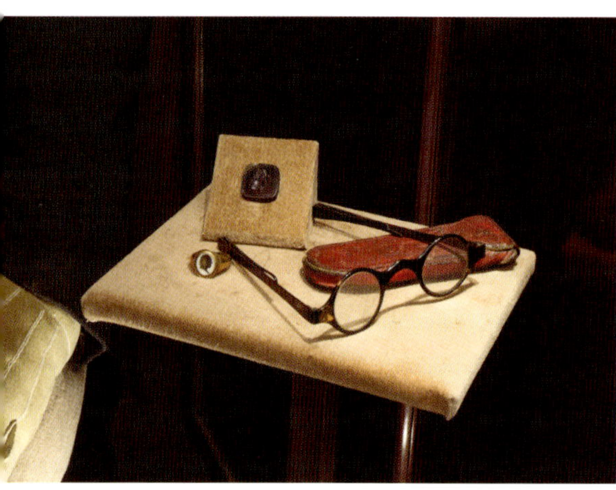

of animals. Above all, he attempted in many writings to think through the political and legislative consequences of this conception of society.

The auto-icon, however, is itself the result of Bentham's application of his own system. Bentham's problem was how, even after death, his remains could be put to beneficial use. His answer had two sides. First his body would be useful if it could instruct students of medicine by being dissected. Bentham pioneered this practice which is common nowadays. Bentham's own body was dissected at a public lecture, which was attended by the famous George Grote whose name is attached to the chair I occupy! Bentham also wanted to establish leaving one's body to science as a general and legal practice. But the second part of Bentham's answer is the auto-icon itself. He thought that having the auto-icon promoted utility. It could be brought out when people were discussing serious issues and would be fun for the participants, fun, of course, being a good thing. But more generally, it seems to me reasonable to conjecture that his purpose is to cause people to rethink their ideas and feelings about death. He is, in effect, saying: do not respond to death by having your body hidden away in the ground, that is to say, being buried. Here is something else you can do with it, something that other cultures already do.

Many of us have adopted Bentham's practice of leaving our bodies to science, but there are not many auto-icons in our society. So that has not caught on as a practice. But the auto-icon also reminds us of Bentham's amazing moral seriousness, an especially inspiring aspect of that great man. In inspiring us in this way, the auto-icon is doing precisely what Bentham wanted it to do.

Jeremy Bentham's auto-icon
1832–3
Bone, wax, wool, straw, wood, leather, copper wire,
glass eyeballs, human hair, 150 x 50 cm
Bentham Collection

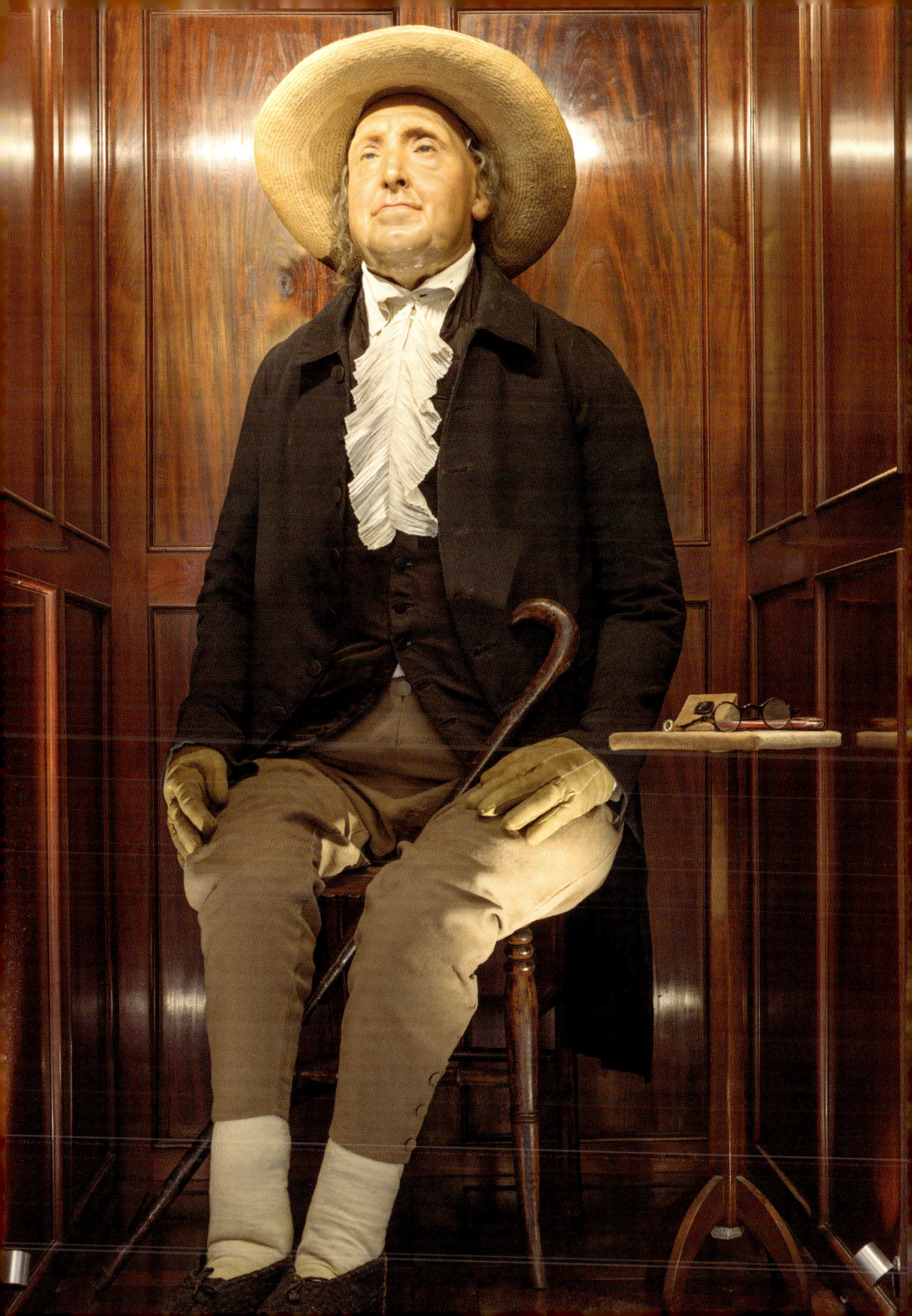

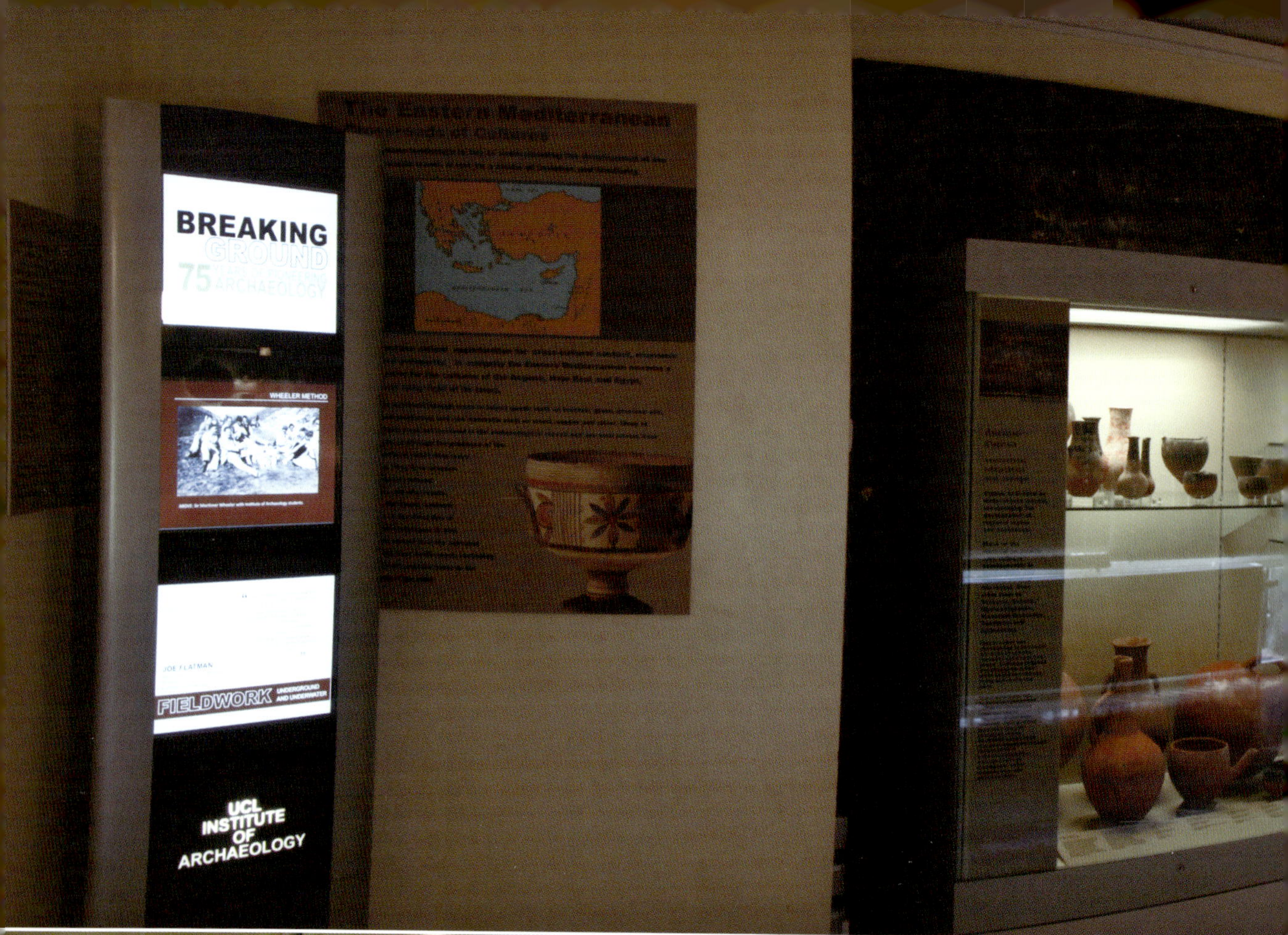

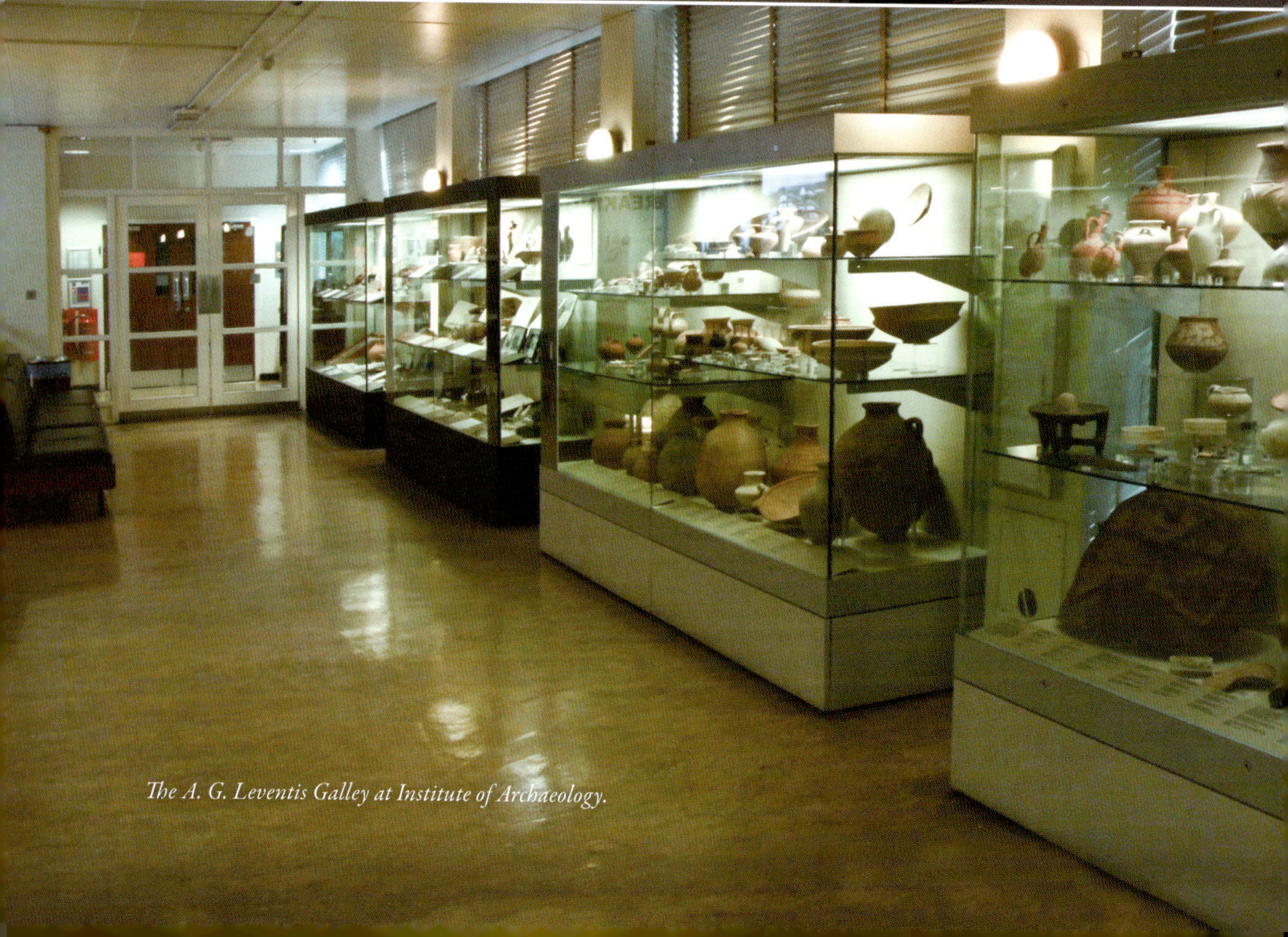

The A. G. Leventis Galley at Institute of Archaeology.

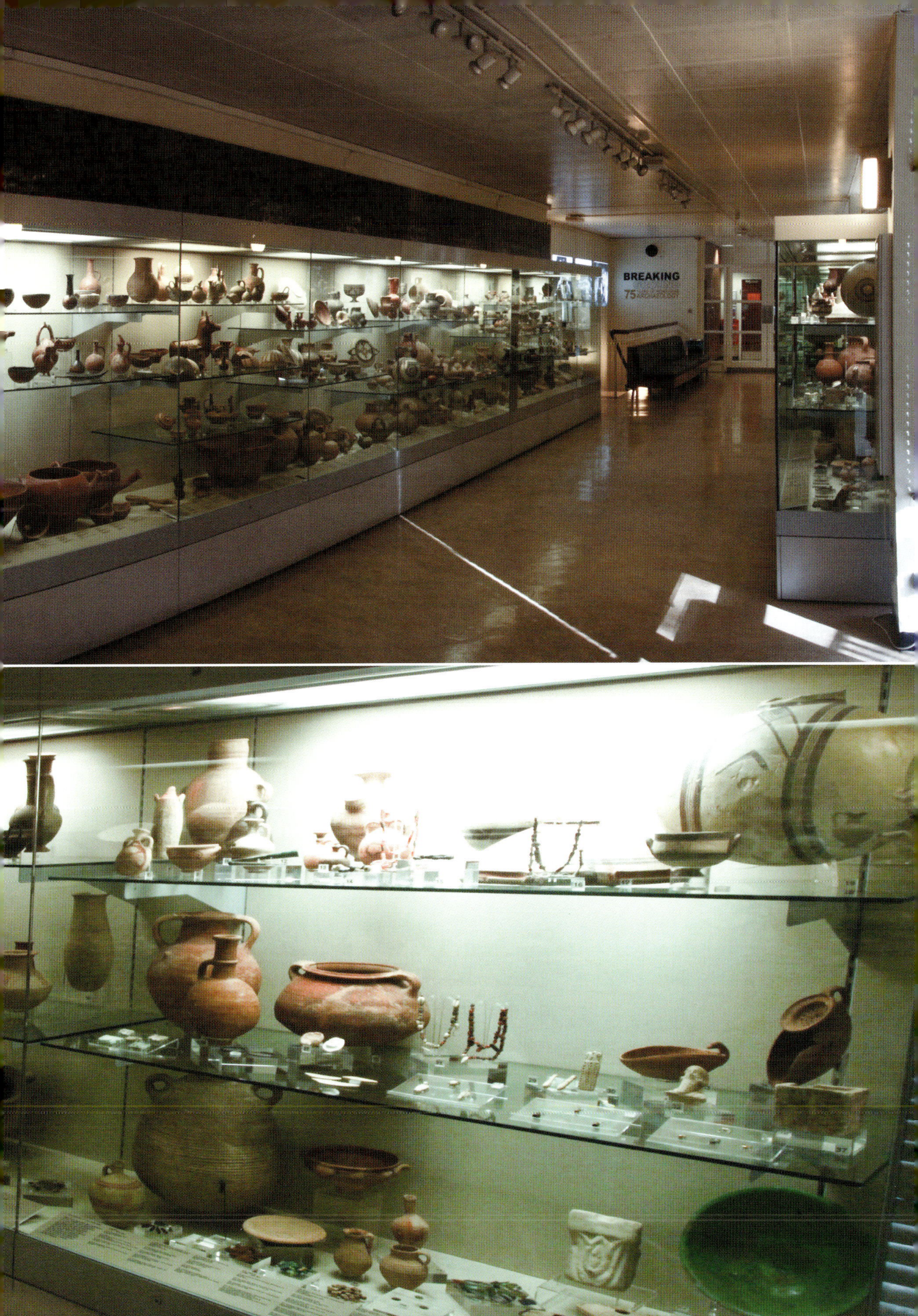

The deed of conveyance of Mery

Professor Stephen Quirke

Institute of Archaeology

EGYPTIAN HIEROGLYPHS ARE more famous, but manuscripts in the cursive flowing handwriting script often answer more of the questions people ask today about social life in ancient times – as exemplified in this legal document, the ancient Egyptian equivalent of a modern will. It is written on papyrus paper, a material which is itself an Egyptian invention of enormous historical impact, as it underpinned the government and administration of the Hellenistic, Roman and early Islamic empires. The earliest known examples of papyrus paper date to about 3000 BC. This document was written over a thousand years later, in the neatly planned 'new town' Hetep-Senusret, a site at the desert edge near the modern village al-Lahun, in Fayoum governorate. It is one of the best preserved among thousands of fragmentary papyri from al-Lahun, the earliest large group surviving, and among the most cited but least seen treasures of UCL Museums; other remarkable finds from this early age of paper include the earliest medical papyrus, the only veterinary papyrus from ancient Egypt, the only two surviving census household entries, and a unique cycle of hymns to the king of Egypt, written out in verses complete with refrain and ditto-spaces.

The document is our only source for the individuals in its story. After date and title *imet-per* 'what is in the house' (meaning 'deed of conveyance'), a man called Mery declares the following:

1. He asks for his son Intef to be appointed to his official position 'controller of the watch' (the person regulating the monthly quota of temple staff); in return Intef should support him as his 'staff of old age' – a technical term for family support, equivalent to a pension.
2. He cancels the preceding deed drawn up for Intef's mother.
3. He grants his estate to the children born to him by a woman called Nebetnennisut – presumably, but not certainly, a second wife.

The document closes with a list of witnesses, partly lost; the title is also written on the other side of the document, in the rectangle that would have been left visible after the document was folded and sealed.

Beside the automatic links to the worlds of the library, the law and book history, the will of Mery conceals an indirect but specific connection to the UCL Faculty of Laws. At al-Lahun in 1889, Petrie worked mainly with Fayoumi Egyptians, but there was one other foreigner: Maurice Amos, son of his late friend Sheldon Amos. Amos senior was UCL professor of Jurisprudence, appointed judge in the Egyptian court of appeal after the 1882 British occupation of Egypt. Maurice followed his father into the legal profession, with a career from inspector in the Ministry of Justice in Egypt, to his own jurisprudence professorship at UCL – not far from Mery.

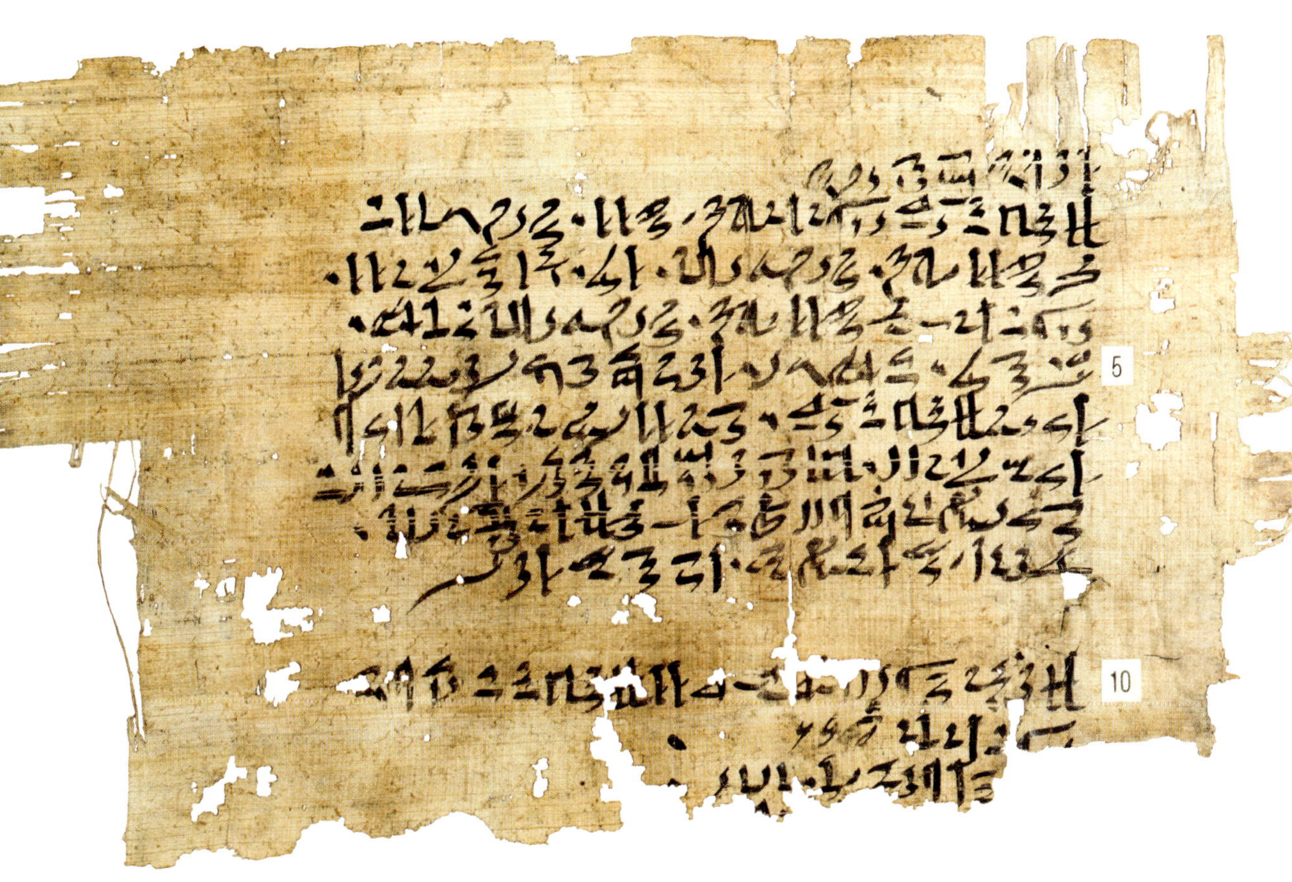

Stanley Spencer's *By the River*

Professor John Mullan

Department of English

I KNOW A FEW PEOPLE who suppose that academics live a better working life than other professionals, their minds of course dwelling on truth and beauty all day. Taking a lunchtime Kit Kat and double espresso in the Housman room directly under Spencer's strange, semi-allegorical painting, I can sometimes allow myself to think that they may be right. This painting and another by Spencer, *Nativity*, dominate the south wall of the room, itself the last redoubt of civilization for UCL academic staff. Spencer's painting was given to UCL in 1963 by Sir Frederick Hooper, former managing director of Schweppes, who had a degree in Botany from UCL. With his success as a businessman, Sir Frederick became a patron of the arts in the 1950s and 1960s and left us this sign of his attachment to UCL. I think that we should be grateful.

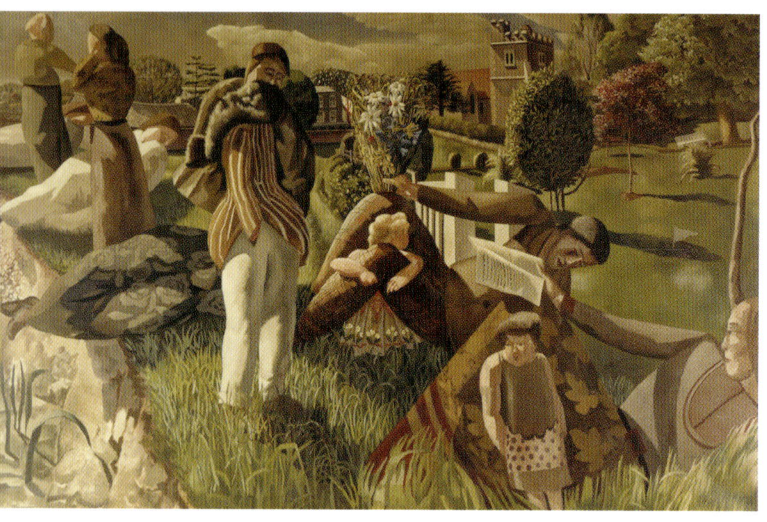

I have spent a good deal of time looking at the painting without quite working it out. Like many of Spencer's paintings, it is a scene from near Cookham, the Berkshire town on the Thames where he lived and worked. It is set in Bellrope Meadow, which is still there if you want to find it, but it is also a place in Spencer's eccentrically Christian imagination. In the background is the squat tower of Holy Trinity Church and a stretch of manicured lawn. In the foreground a collection of mysterious figures are gathered on the narrow bank of the river. Oddly muffled and contorted, they are evidently representative as well as individual.

Spencer was a graduate of the Slade, and the painting hanging next to *By the River*, *Nativity*, was his graduation piece, completed in 1912. It is overtly religious, while *By the River*, painted in 1935, is only implicitly so. If Spencer was not the artist, you might think that these were locals relaxing on a sunny Sunday. But you know that water is the element of baptism and purification, as surely in 20th-century Berkshire as it was in first-century Galilee. Lazily, I have never sought an explanation of the scene, though I know that a Spencer expert could tell me what each of these people is doing, lured to the water's edge for what looks like spiritual communion rather than secular leisure. Spencer was utterly eccentric, but his determination to envisage the events about which he read in the New Testament enacted in his home town is wonderful as well as a little bewildering.

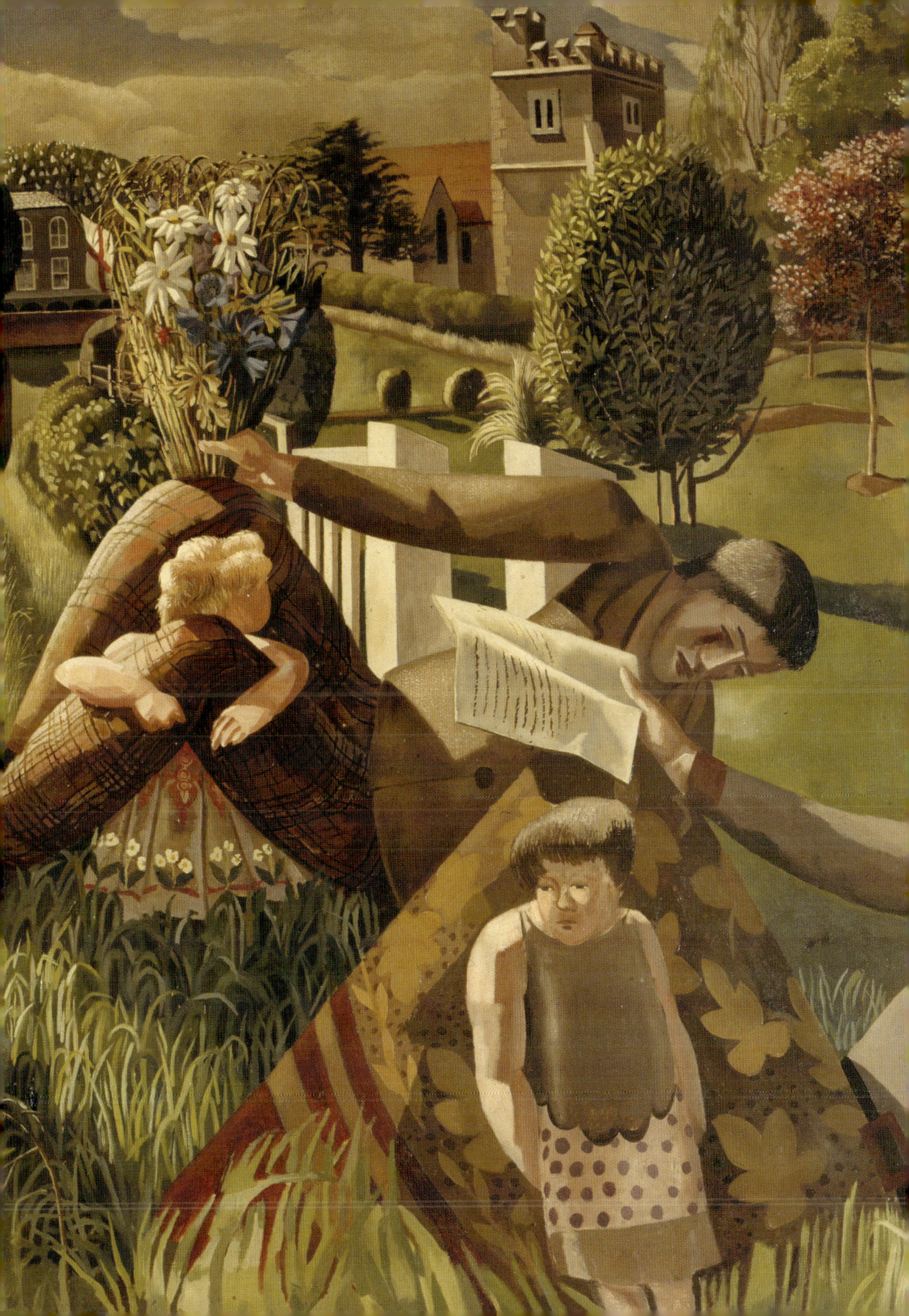

Thalassiodracon

Professor Joe Cain

Department of Science & Technology Studies

HUMANS CAME TO UNDERSTAND the significance of fossils only in the early 19th century: their extraordinary variety and quantity; forms unlike anything seen in the world today; incomprehensibly vast periods of time for layer upon layer to accumulate; arrivals and departures long before we humans entered the frame.

So strange and unexpected were some of these materials that it took generations before disbelief faded and eyes could comprehend what they were seeing. Meanwhile, collectors worked with enterprise and haste to gather and sell, count and measure, compare and classify. Enthusiasts took comfort in technical details: size, shape, arrangement and location. Pieces were given names. Parts were fitted together. Communities were brought back into association. Museums were built to display what had been gathered together.

But science wasn't large enough to hold all the questions people asked about these artefacts. The best minds of the age were drawn into discussions about the meaning of these rocks. Were these objects reality or were they deceptions? Was there deeper meaning, or testimony, in the rocks?

It should come as no surprise that fossils provoked philosophers and theologians from the start. The metaphor of a book took hold. If the history of the world was to be understood as a book, what sort of book had it been? A novel must have a plot. An anthology might have a theme; but equally, it could be a loose array of separate stories. Characters might form a Dickensian multitude, some may be major and others minor, or they might bond together into one glorious epic. The writing of this book may have stopped, a climax reached, or a sequel planned. Our author might be following just an idle whim, or they might have gone on to other things. Perhaps our book had many authors; perhaps, it had only one.

Thalassiodracon is a relic from this great moment in natural philosophy, when hard work led us to worlds of truly unexpected fascination. This is the start of palaeontology – when science, philosophy and devotion sat as siblings together to ponder nature. Imagine chipping away at a stone, then finding something like this. Imagine revealing it piece by piece. Imagine your eyes and hands meeting the first individual discovered of so strange a species, something no other human had ever seen before, let alone dreamt about. Imagine an evening's debate over the meaning of this relic for us and for the book that is the history of our world.

This is a cast. It was purchased as a tool for teaching and study. It was added to the collection so students might measure, count and classify the skeleton found in the rock. It also was added so students might see it through eyes of wonder and discovery.

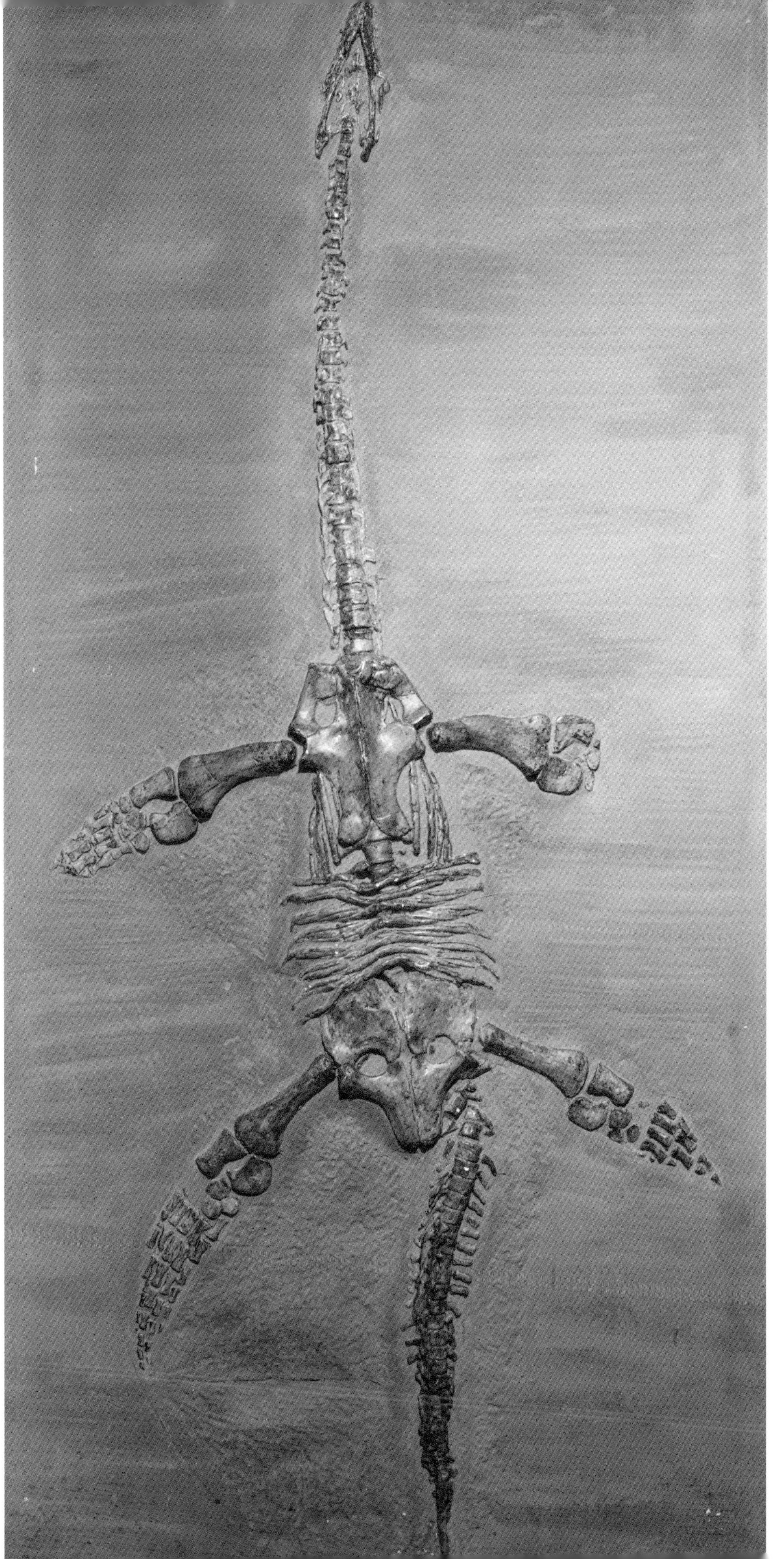

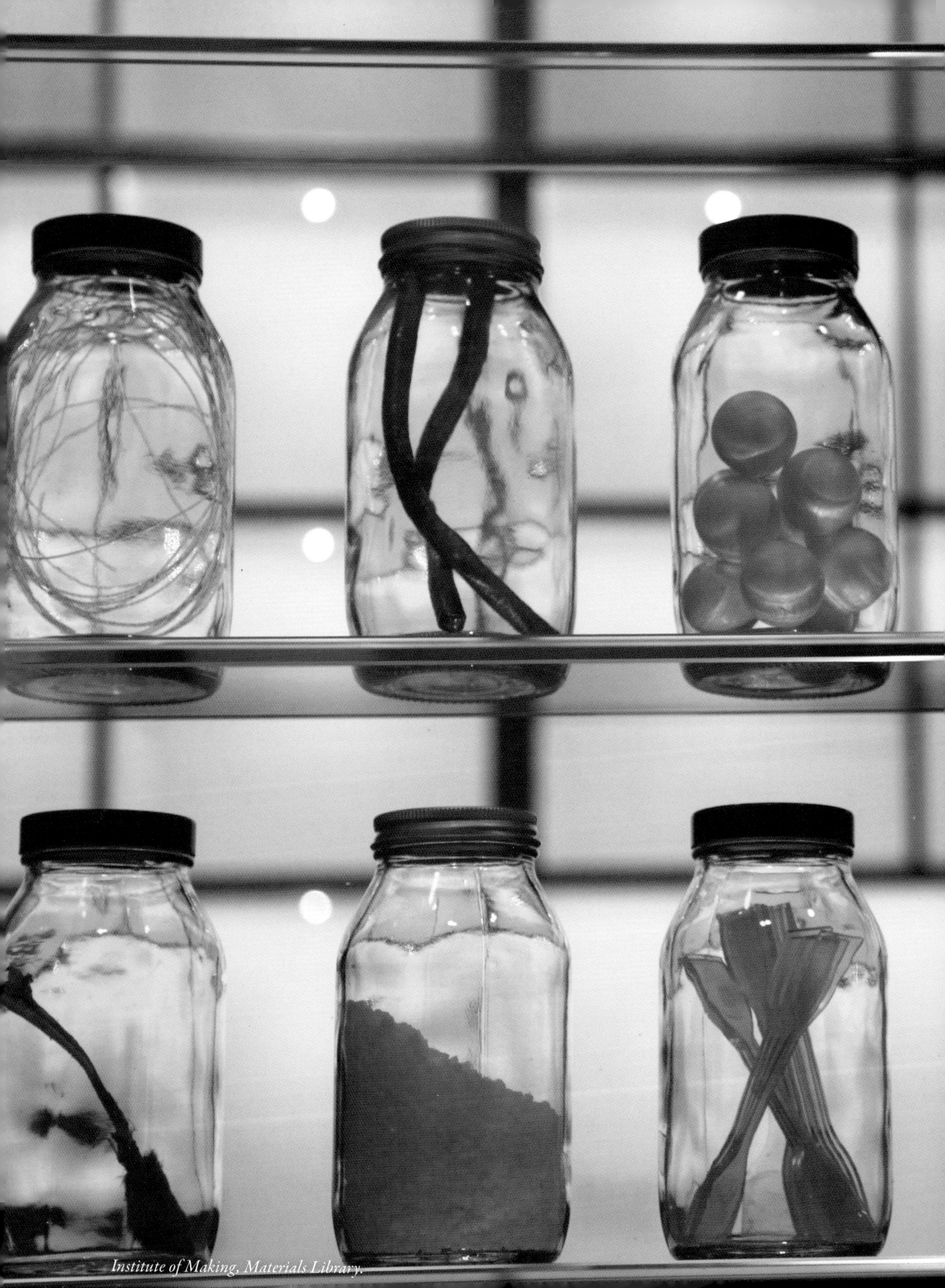
Institute of Making, Materials Library.